Environmental Laws

Environmental Laws

Vierah Hulley

Delve
PUBLISHING
www.delvepublishing.com

Environmental Laws

Vierah Hulley

Delve Publishing
224 Shoreacres Road
Burlington, ON L7L 2H2
Canada
www.delvepublishing.com
Email: orders@arclereducation.com

ISBN: 978-1-77407-755-9 (Hardcover)

ABOUT THE AUTHOR

Vierah Hulley obtained her Ph.D. degree in Geohydrology from the University of the Free State, South Africa, in 2013. Her expertise is in Environmental and Earth Sciences and she has an extensive vocational background in the areas of natural resources management, environmental geochemistry, contaminated site remediation, sustainable development, and spatial analytics. She is the founder and Managing Director of HL Nexus: The Data Imaginarium; a company dedicated to optimizing natural resources management with data science.

TABLE OF CONTENTS

LIST OF FIGURES

LIST OF TABLES

LIST OF ABBREVIATIONS

BPA	bisphenol-A
CAA	clean air act
CAPs	complying assessment procedures
CCS	common compliance scheme
CERCLA	Comprehensive Environmental Response, Compensation and Liberty Act
CERES	coalition for environmentally responsible economies
CITES	convention on international trade in endangered species
CO2	carbon dioxide
CSAPR	cross state air pollution rule
CWA	clean water act
DDT	dichlorodiphenyltrichloroethane
EA	Environment Agency
EAS	environmental agency scheme
EC	European community
EHS	environmental and heritage services
EIA	environmental impact assessment
EMS	environmental management systems
EPA	Environmental Protection Agency
EU	European union
FRG	Federal Republic of Germany
GDR	German Democratic Republic
GHGs	greenhouse gases
HCR	harvest control rule
ICAO	International Civil Aviation Organization
ICC	International Chamber of Commerce
IEMA	International Environmental Management Association
IFC	International Finance Corporation

IPC	integral pollutants control
IPPC	integrated pollutions prevention and control
LAAPC	local authority air pollutant control
LDAR	leak detection and repair
LEV	low emission vehicle
MPO	Metropolitan Planning Organization
NAA	non-attainment area
NAAQS	National Ambient Air Quality Standards
NEPA	National Environments Policy Act
NESHAPs	National Emission Standards for Hazardous Air Pollutants
NMFS	National Marine Fisheries Service
NO2	nitrogen dioxide
NSPS	new source performance standards
NSR	new source review
O3	ozone
ODP	ozone depleting potential
ODS	ozone-depletion substances
OMA	operation monitoring assessment
OPA	operator performance appraisal
OPRA	operator performance as well as risk assessment
PHA	pollution hazard assessment
PSD	prevention of significant deterioration
SEPA	Scottish Environmental Protecting Agency
SIP	state implementation plan
SNAP	significant new alternatives policy
SO2	sulfur dioxide
SWDA	safe drinking water act
TTHMs	trihalomethanes
UNEP	United Nations Environmental Programme
US EPA	The United States Environmental Protection Agency
VALE	voluntary airport low emissions
VOC	volatile organic compounds

PREFACE

Environmental laws are a combination of regulations and agreements that govern human interaction with their environment. In simpler terms, environmental laws can be seen as the laws of addressing environmental problems. Environmental problems vary from a national level to an international level. Examples of environmental problems are animal extinction, pollution, deforestation, climate change, etc. For example, in a nation that faces deforestation, laws may be put in place to protect forests and trees. An individual may be told to cut down a maximum of two trees at a time and plant two trees for every tree that they cut down.

Environmental laws help limit the use of environmental resources and are necessary for the protection of the environment. Environmental laws not only work to promote the value of the environment but also works towards the protection of the environment so that the world population lives within the carrying capacity of the earth. In most cases, environmental laws do not achieve the desired results instantaneously but work gradually to achieve the intended goals. For example, a law passed to reduce deforestation may take years to yield meaningful results.

As stated above, environmental laws vary from a national level to an international level but the general principles of the following laws have been embraced in most parts of the world. USA based laws include, (i) The Clean Air Act (CAA) of 1963 which is a United States federal law aimed at governing air pollution and contamination. (ii) The Endangered Species Act of 1973 which contains regulations to guard endangered species. This Act prohibits unnecessary hunting or killing of animals threatened by extinction. (iii) The Clean Water Act which was passed in 1972. It contains regulations that control discharge of pollutants into water surfaces like streams, rivers, lakes. (iv) The Safe Drinking Water Act (SDWA) of 1974 which protects drinking water quality, focusing on waters actually or possibly developed for drinking consumption, whether from subsurface or underground reserves. (v) The Resource Conservation and Recovery Act (RCRA) of 1976, which allows the EPA to manage hazardous waste from cradle to grave. It includes the production, transportation, treating, storage, and hazardous waste disposal. It also established the framework for managing non-hazardous solid waste. (vi) Comprehensive Environmental

Response Compensation, and Liability Act (CERCLA) of 1980. It offers a federal 'superfund' for cleaning up uncontrolled or discontinued hazardous waste locations, including accidents, spillages, and emergency pollutant releases.

For international treaties, there's also the Montreal Protocol which was agreed upon in 1987 and passed into law in 1983. It is an international agreement designed to guard the ozone layer by eliminating the production and discharge of numerous substances that cause ozone depletion. Some of these substances include chlorofluorocarbons (CFC's), hydrochlorofluorocarbons (HCFC's), hydrobromoflurocarbons (HBFCs), among others. (Chung et al., 2017) Available at: https://content.next.westlaw.com/Document/I466099561c9011e38578f7ccc38dcbee/View/FullText.html?contextData=(sc.Default)&transitionType=Default&firstPage=tru)

Stockholm Convention on Persistent Organic Pollutants (SCPOP) of 2001, it's a global environmental treaty that seeks to remove or limit production and utilization of (POPs) persistent organic pollutants. The Rio Declaration on Environment and Development (RDED) of 1992, this policy is aimed at achieving sustainable development and high life quality for everyone. It recommends that States must reduce and remove unsustainable production patterns and consumption, plus promote conducive demographic policies.

Kyoto Protocol of 1992, it's an international treaty that seeks to reduce greenhouse emissions depending on scientific consensus that global warming is happening and carbon dioxide emissions are responsible. As of now, there are 192 countries signed to the Kyoto Protocol. The Basel Convention of 1992. This is a global treaty developed to limit transportation of hazardous waste across countries, especially to prevent movement of hazardous products from developed to developing nations. All these are important aspects of international law that are ratified by the majority of countries.

In conclusion, the above environmental laws were developed to protect the environment from degradation. Without them, authorities won't be able to punish individuals who handle the environment poorly therefore affecting the health and livelihoods of others. This book seeks to educate the general public and large-scale producers of waste about the existing local and global legislations around environmental pollution, including measures that can be taken to reduce waste emissions.

Chapter 1 discusses the sources and the law making process of environmental law, Chapter 2 about the principles of environmental law, Chapter 3 about the importance of environmental law, Chapter 4 about the clean water act, Chapter 5 about the clean air act (CAA), Chapter 6 about the endangered species act,

Chapter 7 about the safe drinking water act, Chapter 8 about the comprehensive environmental response, compensation, and liberty act (CERCLA), Chapter 9 about the resource conservation and recovery act, and Chapter 10 about the problems/challenges in implementation of environmental laws.

1

The Sources and the Environmental Law-Making Process

CONTENTS

1.1. INTRODUCTION

Environmental law can be defined as the group of laws, regulations, directive, policies, treaties, and common law governing how people interrelate with their natural environment. The primary purpose of environmental law is protecting ecosystem resources and creating policies for ways individuals may utilize natural resources in a sustainable manner. These policies not only aim at safeguarding the ecosystem from destruction, but further regulate those that can utilize nature's assets and under what conditions (Figure 1.1).

Figure 1.1: Environmental law protects nature from destructive practices.

Source: https://oikon.hr/our-departments/department-of-environmental-law-policy-and-economics/.

Laws have the capacity to moderate pollution, including other factors such as mineral harvesting, forest protection, and flora and fauna populations. Nevertheless, environmental laws are comparatively new in history. Policymakers started to enact conservational measures in around 20th century. In particular, environmental campaigns began gathering pace during the 60s, with a number of conservational laws and regulations in existence today developed around the period (Fisher, 2017), particularly in the USA, Europe, Australia, Japan, and New Zealand. Rapid re-industrialization in Japan post World War II, led to the contamination of human food chains such as the case in Minimata where a significant number of the population suffered from mercury poisoning after eating fish contaminated with industrial wastes. Significant environmental disasters such as these led the

Japanese government to enact the world's first overarching pollution-control policy in 1967, the Basic Law of Environmental Pollution Control.

Original environment legislations were concerned with mere nuisance matters. Cases where an individual's use of their resources inhibits another individual's use of their personal property, courts may come in to break the nuisance. These nuisance laws mainly developed through regular law decisions made in courts. They protect property owners from having another individual or organization invade on their rights for enjoying their own property. Nevertheless, early environmental laws were not concerned with safeguarding the environment entirely. They also didn't provide standing for an individual to sue the emitter if they weren't individually affected by other individuals' actions (Fisher, 2017).

The first federal environment law to be enacted was Rivers and Harbor Decree of 1889. Later, Clean Waters Act reviewed a large part of Harbor Decree. Finally, The Act criminalized introduction of waste products into maneuverable waterways without any valid license. Laws further made it criminal to modify the harbor or water-front, or alternatively change the watercourse through fill up or excavation. Other considerable aspects of civic environment laws are: Clean Water Act (CWA), the CAA, Compensating, and Liabilities Act, Comprehensive Environmental Response, Threatened Species Act, Resources Conservation and Recovery, including the National Environmental Policies Act (Ferrey, 2010).

With environmental laws progressing, lawmakers are continually addressing the matter with greater focus. People don't have to get individually distressed to present any claim for stopping ecological pollution. Lawful variations have further inspired innovative legal activities, such as Scenic Hudson Protective Convention vs. Federal Power Convention, preventing the power station from working. Courts determined that a collection of residents after appearing in the court to contest a developmental project, due to environmental concerns, won the case (Sancin, 2012).

The 1970's and 1980's saw a significant evolution of international environmental laws. In 1971, 34 countries adopted the Convention on Wetlands of International Importance Especially as Waterfowl Habitat, generally known as the Ramsar Convention. The United Nations Environmental Programme (UNEP) was established as the principal international environmental organization after the 1972 United Nations Conference on the Human Environment. The early 1970's also led the

European Community (EC) to becoming the forefront in the promulgation of environmental laws that seek to balance economic development with the need to protect the environment.

1.2. WHAT EXACTLY DO ENVIRONMENTAL LAWS MANAGE?

These laws encompass a vast range of topics that including the following areas:

Environmental laws can be divided into the following categories:

- Command and control. These include legislation that:
- Control Pollution (such as waste management, air quality standards).
- Resource management (such as wildlife management legislation, forestry laws).
- Environmental assessment mandates-protect the environment indirectly by increasing the quantity and quality of publicly available information on the environmental consequences of contemplated actions.
- Re-active management of impacts such as the environmental remediation laws.
- Economic Instruments: These create incentives to protect the environment. Examples include: pollution taxes, landfill taxes, subsidies for cleaner production and technologies.
- **Mineral Resources:** The objective of this law is supervising the managing and exploiting of minerals and all activities associated with mining operations. Often mineral resources are state property, and legal entities or individuals intending to engage in mining activities will need a mineral resource approval issued by a relevant authority. In order to promote the management and exploiting of mineral resources, different types of permits should be provided. The Mineral Reserves (Sustainable Development) (Mining) Regulations of 2019 came into effect on July 1, 2019. These regulations replace the former regulations, which were repealed on June 30, 2019.

The regulation has made many improvements to simplify and update the regulations, plus has developed an open work plan and recovery plan requirements to confidently manage the risks related to mining and mineral exploitation. These policies are formulated through the public comments provided by the "Regulation Impact Statement" procedure.

- **Forest Resources:** Policies and regulations are developed from a collaborative process between stakeholders and sanctioned and published by registered institutions. National and county governments have passed regulations and other kinds of legislation for forest preservation. Since we all recognize this, there's need for striving to achieve similar or even common goals, particularly at the nationwide level. These goals include: improving efficiency and effectiveness, improving competitiveness, promoting trade, improving citizen welfare and safeguarding the environment. Therefore, it is not surprising that working with government agencies to formulate community-oriented policies and laws.
- **Biodiversity Management:** Different kinds of land are found in forests, parks, and other environments, supporting animal and plant communities. Environmental laws have helped to manage the diversity or biodiversity of the plants and wildlife found there. The management of biodiversity conservation needs that important biodiversity areas be reserved and incorporated into every area of management. Through proper management of land and vegetation, a variety of species will benefit. (United Nations Environment Programme, 2020) Available at: https://wedocs.unep.org/bitstream/handle/20.500.11822/25655/LawBiodiversity_Strategies.pdf?sequence=1&isAllowed=y
- **Impact Assessments:** Environmental Impact Assessment or (EIA) is the process of assessing the possible environmental effect of a suggested project or development program, taking into effect the related socioeconomic, cultural, and human wellbeing impacts, including good and bad. UNEP describes EIA as an instrument used to determine the ecological, social, and economic effects of a project before making a decision.

It seeks to predict environmental effects in the early stages of project planning and development, determine ways and techniques for reducing adverse impacts, adjust projects to fit the local environment, as well as provide forecasters with choices and predictions. Through using EIA, both environmental and economic advantages can be attained, like reduced cost and duration of project application and design, prevented treatment/cleanup rates and effects of laws and legislations.

- **Mining Resources:** International treaties recognize the need for cooperation in the management of the environment and natural products, and coordinated measures should be considered to promote cooperation within the community for effective management and sustainable use of natural resources, so that signatories are mutually beneficial. Regarding the managing of mineral resources, it was agreed that the shared mineral resources must be effectively developed and sustainably used.

Various partner countries have made efforts to formulate mineral policies and mining systems. Therefore, the regulator plans should be preparing a list of mineral or mining policies for countries within its jurisdiction. A feasibility evaluation has been conducted on the mineral resources in different areas. This value-added potential can add value to the local area in order to create local employment opportunities and guarantee that the local community gets the most benefit from the mineral assets in the given area.

Environmental laws cover a vast range of subjects in these fields:

- **Water Resources:** Ecological regulations safeguard aquatic systems from contamination. Besides, they might decide individuals that utilize water resources including the best ways of handling probable problems such as treatment of wastewater and management of surface runoff. Water laws include managing quantity of water as well, for example, the aquatic reserve. (Food and Agriculture Organization, 2009) Available at: http://extwprlegs1.fao.org/docs/pdf/tan96340.pdf
- **Waste Management:** Public refuse, hazardous, and toxic wastes as well as nuclear wastes are classified and managed appropriately to minimize negative human health and ecological impacts.
- Air Quality Standards: These policies safeguard the atmosphere from unnecessary contamination and might cover procedures for protecting the atmosphere from conditions such as ozone degradation.
- **Environmental Cleanup/Remediation:** In contrast to pollution control laws that are proactive, regulations on environmental cleanup/remediation (also referred to as contaminant cleanup) addresses managing contamination following its occurrence. Legislation might cover procedures for site assessments, remediation, site reuse, public consultation and felony sentencing for contaminators.

- **Chemical Safety:** These guidelines control the potential pollutant and govern the use of the chemicals such as regulating pesticide usage and banning of harmful compounds such as the use of ozone-depleting substances in consumer products or the use of bisphenol-A (BPA) in plastic bottles.
- **Game Hunting and Angling:** The policies might control and safeguard wildlife numbers. Legislators decide those capable of hunting, fishing, or how these practices are regulated.

The conservational rules or principles offer the foundation for guarding the ecosystem from potential damage; short of these, it means things shall get in bad condition. Conservational rules offer the control for the haphazard discharge of harmful compounds within the ecosystem or aquatic channels, apart from helping to avoid devastation to the globe's natural courses, habitations as well as life forms.

With increased understanding on the impacts of pollution, so have the policies and regulations attempting to manage those actions which make them to rise. Increasing laws for protecting the ecosystem are constantly getting ratified by lawmakers, having firmer discharge parameters to reach. The comparatively current introduction of industry-oriented tools, like harmful gas release dealing, has offered a new aspect towards controlling pollutants by organizations (Fisher, 2017).

Moreover, organizations get a lawful and ethical obligation towards conforming with ecological legislations plus guidelines; since regulation obedience serves as the community's warrant for operation. Considering the complex nature of lawful policies concerning ecological safeguarding, still, it's a considerable setback to some organizations.

Among the first challenges comes from discovering the rules plus guidelines that really apply; closely associated with the requirement to comprehend their applicability, including factors that should be performed to fulfill and keep obedience on a continual way. It requires time as well as resources, alongside solid managing, in case it's to be performed properly. Ecological managing structures provide an instrument that organizations may easily control, amid other factors, their supervision submission and enhance supervisory performance (Ferrey, 2010; Sancin, 2012; Koivurova, 2013)

From face value, organizations having Environmental Management Systems (EMS) possess improved monitoring standards in comparison to the ones lacking these solid structures. By definition, an environmental management system refers to an inventory or database that combines procedures and protocols for training staff, condensing, monitoring, and reporting unique environmental performance details to both interior and exterior institutional stakeholders. It allows organizations to reduce their environmental impacts or increase operational efficiency.

Nevertheless, there's been a lot mentioned in the environmental media regarding failure by certain agencies to attain regulation compliance on a continual measure, despite the EMS being licensed as ISO (14001) and listed to EMAS. Lots of disapproval has further come from certifiers; who are UKAS, and to the credit, followed activities towards enhancing how it licenses certification agencies, besides IEMA sought to assess its effectiveness back in 2006.The International Environmental Management Association (IEMA) is an international alliance of environmental and sustainability professionals.

Currently, bodies such as REMAS LIFE funding program, a collaboration among the UK Environmental Association, Scottish Environmental Protecting Agency (SEPA), Ireland EPA, plus IEMA, is looking to institute values of EMS towards regulation agencies plus the outcomes are keenly anticipated. This is a public-private partnership of UK environmental regulators working together with professionals/consultants belonging to the IEMA to promote the use of EMS.

Each of the activities shall assist to improve the efficiency of EMS towards controlling regulation observance and routine. Although an EMS can't assure lawful obedience, for efficacy it must function like an instrument such that organizations recognize their compliance condition plus remedial or preventive activity that may be swiftly applied. Supervisors, certifiers, and citizens must not be in a position to find aspects of legal non-compliance (Ferrey, 2010). (American Society for Quality, 2020) Available at: https://asq.org/quality-resources/environmental-management-system

1.3. ENVIRONMENTAL LAW SOURCES AND NATIONAL REGULATORY BODIES (THE UNITED STATES EPA)

1.3.1. Sources of Environmental Laws

Environmental laws have been developed from various places. National governments protect the environment based on their governing constitutional obligations and within the statutory powers to promote general well-being, manage public natural resources (air, water, land) and regulate commerce. They may also accept additional duties to protect the environment by entering bilateral and/or multilateral treaties, which may contain specific obligations. Administrative authorities may promulgate environmental regulations and permits, which may contain compliance requirements related to reporting, monitoring, and civil and/or criminal actions.

In the United States, many federal regulations are derived from the United States Environmental Protection Agency (US EPA). In specific, Title 40 of the Federal Regulations Code has various environmental regulations developed by the US EPA. One common case that affirms the EPA's extensive powers is the Calvert Cliff Coordination Commission versus. American Atomic Energies Commission. US EPA utilizes its Department of Enforcement and Complying Assurance to impose its regulations. Some other federal organizations which manage the environment are U.S. Fishes and Wildlife Service, National Park Services, U.S. Forests Department and Agency of Land Management (Sancin, 2012).

International agreements are a big part of legislation. In particular, ecological regulation is a subject of global discussion and debate. Governments around the world grapple with the concept of how they can effectively control pollution and utilize natural resources from a worldwide scale. Since environmental regulations may be a particularly hot topic among regional, state, and national administrations, environmental management and usage of natural resources still remain to be controversial subjects of discussion between governments (Fisher, 2017).

Among the first global summits on environmental matters was the 1972 UN Conference on Human Environment. It was then followed by the 1992 UN Earth Summit which came up with the renowned Rio Declaration. In this charter, signatory states mentioned that humans must develop the globe in a way which fulfills developmental and environmental requirements of current and future generations.

While there are federal policies and international agreements concerning environmental law, there's still a great part played by regional and local/municipal government concerning the environment. Regional or local governments may have unique agencies that can help manage and control natural resources within the area. These local agencies establish regulations concerning aspects such as how people may hunt or fish. (Fisher, 2017).

Environmental laws are rather controversial and vary from one region to another. Policymakers should discuss the necessity, impartiality, and affordability of environmental regulations. However, undertaking a cost-benefit analysis of environmental regulations is filled with shortfalls and multiple variables, making the estimation of such extremely difficult. Given the variables and the uncertainties with such an estimate of a regulation, people can still have a great amount of disagreement concerning the advantages of regulation.

1.3.1.1. National Environmental Law

The four main sources of national environmental law include:

Constitutional Law: The constitutions of many countries include provisions establishing environmental rights, or government duties on environmental protection and management of natural resources. For example, environmental protection is enshrined in the German-German Grundgesetz ("Basic Law") which states that the government must protect "future generations the natural foundations of life."

Section 24 of the South African Constitution (Republic of South Africa, 1996) enshrines environmental rights in South Africa:

"Everyone has the right,

- to an environment that is not harmful to their health or well-being; and
- to have the environment protected, for the benefit of present and future generations, through reasonable legislative and other measures that:
    - prevent pollution and ecological degradation;
    - promote conservation; and
    - secure ecologically sustainable development and use of natural resources while promoting justifiable economic and social development."

In South America, the constitution of Argentina entitles any person to initiate an action for environmental protection while the Chilean constitution provides for the "right to live in an environment free from contamination."

Environmental Statutes: The bulk of environmental laws are developed by government agencies charged with environmental protection. Legislative texts often establish general environmental policy, supplemented by specific laws and administrative regulations. Environmental statutes use generally applied principles and techniques to protect the environment, such as EIAs, licensing, the setting of emission, remediation, or water quality standards; often regulating a singular environmental medium such as biological diversity, waste or water management, soil, or air. There remains an increasing trend to develop a comprehensive approach to environmental protection by applying methods to integrate pollution control and prevention. Environmental statutes also respond to a specific environmental concern for that particular countries. Examples of these may include specific environmental requirements for the oil and gas industry or for the mining and metallurgical processing industries.

Administrative Regulations: A good part of environmental laws enforcement happens through administrative law. The USEPA may examine a violation or bring an administrative aspect to their particular officials. Attorneys represent the EPA during these hearings, plus they represent the individuals and businesses accused of rules violation. Entities who are found liable of breaking the rules may petition the decision with established courts. Many violations are civil offenses, however, there are also criminality penalties for intense offenders.

Industry Standards and Codes of Conduct: Transnational corporations and other business enterprises shall carry out their activities in accordance with national laws, regulations, administrative practices and policies relating to the preservation of the environment of the countries in which they operate, as well as in accordance with relevant international agreements, principles, objectives, responsibilities, and standards with regard to the environment as well as human rights, public health and safety, bioethics, and the precautionary principle, and shall generally conduct their activities in a manner contributing to the wider goal of sustainable development.

- The UN Sub-Commission on Human Rights, 2003

Globalization has led to an increase in the drafting of guidelines and codes of conduct within industry and multinational enterprises. Examples of this includes the World Industry Council for the Environment, the International Finance Corporation (IFC) Sustainability Policies and Framework, the FAO International Code of Conduct on the Distribution and Use of Pesticides, the Responsible Care Initiative of the Chemical Manufacturers Association, the Coalition for Environmentally Responsible Economies (CERES)/Valdez Principles, the International Chamber of Commerce (ICC) Business Charter on Sustainable Development, the Royal Dutch/Shell Group Statement of General Business Principles.

1.3.1.2. International Environmental Law

International law is the reigning body of law invoked by international tribunals and considered in matters of international relations between nations. When international law has been ratified by a country and made binding, its ranking compared to a nations legal sources depends on the stipulation in the constitution. Increasing environmental awareness and activism in the mid-twentieth century resulted in the development of international environmental law as a subset of international law. The historical evolution of international environmental law can be divided into three phases, separated by two of the most relevant international conferences held so far: (i) the "traditional era" which is characterized by the period prior to the 1972 United Nations Stockholm Conference on the Human Environment, (ii) the "modern era" which characterizes the period between the Stockholm Conference and the 1992 United Nations Conference on Environment and Development in Rio de Janeiro, commonly referred to the Rio Conference, and (iii) the "post-modern era" which marks the developments after the Rio Conference to present.

International environmental law is produced through the interplay of international organizations, conferences, diplomacy, codification, and progressive development, and international courts, and the interplay of treaties, non-binding declarations or resolutions, and customary international law. The United Nations and other key international agencies lead the role in setting the agenda for international environmental laws. These organization also provide essential tools for the process such as international negotiating forums for building consensus and package diplomacy, and expertise.

1.3.2. Environmental Justice

Environmental justice refers to the just treatment and meaningful participation of everyone in formulating, implementing, and enforcing environmental laws, principles, and policies, regardless of race, skin tone, nationality or income. Fair treatment in environmental justice means that no one is responsible for the excessive proportion of negative environmental effects caused by industrial, municipal or commercial operations and enforcement of federal, statewide, and local laws; policies and regulations. Appropriate participation requires all people to have practical access to decision-makers, plus all communities have the capacity to make knowledgeable decisions and take active steps to fight for environmental justice for their own good.

The environmental justice campaign was began by individuals (mainly people of color) who tried to solve the inequality of environmental protection within the community. Professor Robert Bullard writes: "Whether it's intentional design or institutional neglect, people of color in urban slums are located in rural poor areas. The civil rights crusade of the 1960s rang the environmental alarm that threatened the health of their families, communities, and even themselves.

1.3.3. Environmental Lawyers

Environmental law refers to a broad legal field that deals with a variety of topics related to the environment, such as water and air grade, hazardous waste, agriculture, wetlands, species protection, waste management, eco-initiatives, sustainable strategies biodiversity, and alternative energies. As sustainability grows to be a universal priority, attorneys who can advise customers on green principles and sustainability matters are needed. It's projected that climate change and global warming regulations will surge the legal task of environmental attorneys in the next few years.

An environmental lawyer is committed to representing clients on legal matters such as clean technology, climate change laws, water-based laws, and managing of lands with indigenous ownership and other communal lands. Other key areas include environmental rights, universal environmental law, ocean law, and international resource legislation. Environmental lawyers furthermore work in the fields of climate change, renewable energy (a growth area for global companies), and sustainability. Environmental law is a large and multifaceted area of legal practice. Successful environmental lawyers must be able to efficiently communicate verbal and written documentation.

1.3.3.1. Becoming an Environmental Lawyer

To professionally practice environmental law, there needs to be some considerable magnitude of diplomacy. You may have to discuss a local or municipal government concerning the advantages or possible risks of regulation. Furthermore, environmental attorneys work in both public as well as private practice. They may be employed by big law firms or operate as in-house legal counsel, however, they might also perfectly represent people in small business or solo practice on environmental matters (Ferrey, 2010).

The USEPA moreover needs attorneys to act as administrative adjudicators that hear claims of rule violations. Companies depend on environmental attorneys to help them comprehend and comply with established regulations. They refer back to their in-house attorneys for guidance, though some companies prefer contracting independent lawyers. Even attorneys in private practice might have some few clients for their career in environmental compliance. There could be a considerable amount of litigation that's related to the practice of environmental law, which provides substantial opportunities for environmental legal counsel.

Attorneys must represent their clientele in administrative hearings including formal courts, they should discuss possible resolutions with EPA lawyers as well. To prosper in environmental law, the lawyer should be effective at negotiations. Oratory skills as well as trial advocacy abilities also are important for representing clienteles at hearings, including depositions and even political bodies (Ferrey, 2010).

Environmental attorneys work all over the U.S and the globe. Working for a global company might mean worldwide travel or permanent universal assignment. Businesses should comply with environmental matters throughout the U.S., therefore, they rely on attorneys throughout the region. Additionally, persons who depend on lawyers for bringing cases to safeguard them from ecological pollution or nuisance need attorneys to operate throughout the country. From a local level, attorneys may represent people when facing criminal or civil charges of hunting or fishing violations.

1.3.2.2. Role of an Environmental Lawyer

Environmental law is essential for both people and companies. For attorneys representing businesses, they offer a crucial service which allows businesses to do business in a rather lawful means. Attorneys working for the Environmental Protection Agency (EPA) always put the public interest first. Additionally, attorneys help people to protect their private peace, and conveniently when they're ever distressed by rule breakers (Louka, 2006).

Attorneys in the sector have an opportunity of influencing regional, state, nationwide, and global policies. For attorneys with individual goals of influencing environmental strategies, there are aspects of law that provide this particular opportunity. Besides, where the law and environment interconnect, environmental attorneys contend with whether or how lawmakers need to manage the environment. Whenever such decisions are made, attorneys help clients to apply regulations in a convenient way. Attorneys for government agencies or individuals offer a crucial public service in developing, enforcing, and even challenging environmental laws or regulations throughout the U.S. (Louka, 2006).

1.4. COMPLIANCE EVALUATION

The goal of rule observance evaluation is checking the agreement of structured organizations with appropriate conservational law requirements, further consisting of openly implemented protocols, stipulations in licenses and other lawful obligations which are practical; plus, monitoring the effect of controlled organizations on the ecosystem to decide if additional examination or application activity is needed to guarantee compliance with ecological legal requirements. It's the role of operatives to supervise their actions for achieving lawful agreement, plus it's also burden on the supervisor to determine if the operative is fully compliant with law necessities and determine if additional activity is needed (Bodansky, 2010).

The operative is considered to be conforming when there is minimal evidence for failure to achieve licensing (or alternative) stipulations. Supervisors expect individuals to perform regular assessments testing if they're in conformity. Similarly, assessments must be risk-oriented, based risks presented by functional actions. The methodical technique for compliance managing must give signal of how ecological law requirements are evaluated and achieved, plus detect any legal noncompliance.

Supervisors normally scrutinize the evidence given by operators to help in determining if permit stipulations, or alternative needs have been achieved. Furthermore, the rate of regulation evaluations would be determined by available proof and practical criteria of performance. Besides, there could be conditions in which the operative isn't wholly compliant with few or multiple circumstances mentioned in the license, or some other legal stipulations. The operator must instantly inform regulators of noncompliance, plus consent with supervisors the activities required to return back to compliance apart from mitigating damage to the environment (Bodansky, 2010).

Compliance Obligations: Compliance has been described by the European Union (EU) as "Implementing plus Enforcing of Environment Laws," or complete execution of ecosystem necessities. Compliance happens where certain obligations are achieved plus necessary modifications made. Compliance assessment or regulation is necessary and refers to the general approach taken to review compliance. Monitoring furthermore can be done either by some operator (subject to legislation), or the regulator enforcing legislation (Figure 1.2).

Figure 1.2: Everyone should follow protection policies for preserving the planet.

Source: http://www.courses.knect365-learning.com/event/environmental-law-and-practice-distance-learning-llm-courses-de-montfort-university-dmu/user.

The main goal of environmental regulation is protecting people's health plus general ecosystem from damage around the framework of viable development. Conventionally, ecological regulations have encompassed environment-based media concerning (land, water, and air), alongside wildlife safeguarding and preservation. It has concentrated on controlling of pollution discharges and maintenance plus enhancing of water standards and refuse managing (Bodansky, 2010).

Starting from the 90s, a rather incorporated technique has been tried through different broadcasting networks with the (IPC) Integrated Pollutants Control based in UK, Scotland, and Wales, much lately, the Integrated Pollutions Prevention and Control system or (IPPC). Going into the future, basic conservational laws are highly likely to be adopted in different areas such as:

- Managing environmental effects of products and related activities and practices (like manufacturing and transportation) by embracing a lifecycle approach;
- Effective usage of resources being an aspect of the journey towards viable usage, as well as generation;

- Maintaining and enhancing wildlife diversity; and
- Business and collective social responsibility.

Furthermore, regulation offers environment officials certain obligations and influences. These responsibilities are among the factors supervisors should perform and thus seem to be prescriptive; case in point, the responsibility of determining or issuing environment licenses. Powers are normally flexible on how they are utilized; for instance, the ability to enter buildings or apply proper execution actions (Bodansky, 2010).

When powers have already been applied, take for instance, entering a premise, there could be duties which are applicable like prohibiting activities causing damage to the ecosystem. Additionally, regulators have access to multiple implements at their disposal for application of responsibilities or powers varying from availing counsel plus direction over to formal applications as well as trial. (Regional Environmental Center, 2008) Available at: https://ec.europa.eu/environment/archives/enlarg/handbook/handbook.pdf

Additionally, compliance is measured applying one or several of these kinds of assessment:

- **Continual Evaluation:** It happens where regular checking of systems records information constantly in actual moment for alarm reasons, or through archive means for successive evaluation. For this example, gear is fitted for observing essential vapor releases and stack discharges, usually at complex venues. Nevertheless, not every aspect of the site's actions are able to get monitored this way, meaning thus continual evaluation, when done professionally, seems to be discerning.
- **Retrospection Evaluation:** Episodic and yearly evaluations depend on a reflective evaluation of compliance whereby evidence is determined by regulators for confirming that the entity has been in obedience. EMS assessments and documentation may offer this proof.
- **Periodic Evaluation:** It's the most regular type of regulation evaluation. If applied as an aspect of the risk-based technique, this form will make optimal use of worker assets. There are a total of three subsets available which include: (i) Site-based evaluation: regulation officials visit a venue and perform either a thorough evaluation (e.g., legal compliance auditing), or targeted

inspection (e.g., check assessment) depending on predicated evaluation criteria. (ii) Report-oriented evaluation: regulation officials obtain and inspect data returns, research outcomes, and program feedbacks back in their headquarters, amid site visits. (iii) Condition-based evaluation: regulation officer's reply to events reported to them by operators or secondarily as an objection by some members of public. This reply may sometimes comprise of physical presence at the venue of the incident.

Yearly evaluation: here supervisors prepare yearly data and information meant for input into yearly plans and reports. They comprise of summary information and it's essential to review compliance on hindsight to attain fair resolutions.

1.4.1. Monitoring

Basically, two kinds of monitoring are done or needed by supervisors; which include: - assessment of regulation alongside environmental legal conditions, for instance, permit stipulations: plus, finally monitoring of environmental qualities, for instanee, meeting environmental quality measures. Assessment may be performed in various ways (that is, continually, occasionally, or remotely) plus by operators through self-assessment and regulators. The specific blend of monitoring shall rely on regulatory measures and resources, operator's competency, third party participation, and accessible methods or innovation. It's vital that organizations provide monitoring results which are valid, dependable, precise, and appropriate (Ferrey, 2010).

Monitoring requires the appropriate use of convenient methods, protocols, solutions, and equipment, skillful, and qualified staffs, effectual planning, quality assurance as well as quality control. In the UK, their Environmental Agency has developed a Monitoring Certification Scheme (MCERTS) for improving and ensuring the quality of environmental data, aimed at environmental permit holders.

In Scotland, the national environmental regulatory group, SEPA is presently looking for MCERTS recognition for workers involved in physical stack discharges assessment. Environmental and Heritage Services (EHS) from Northern Ireland is also encouraging operators to utilize MCERTS-certified continuous discharge monitors and get manual spot monitoring functions done by MCERTS credited as per the Environment Agency's (EAs) Operation Monitoring Assessment (OMA) System (Ferrey, 2010).

The EA has introduced this program to reinforce its auditing of operators' individual-monitoring arrangements. Furthermore, OMA is presently used for monitoring of discharges to the atmosphere from manufacturing activities. This program has been suggested for extending OMA to alternative media sources and regulatory systems. The EA utilizes OMA scheme for: assessing operator's self-evaluation (including monitoring done on behalf of workers by contractors) through a consistent and clear approach; contributing to the Agency's general sampling and check evaluation program; apart from providing the driver for essential improvements.

1.4.2. Enforcement

Enforcement refers to stringent action taken by the regulators to ensure organizations fulfill legal requirements of environs licenses, and directly implemented laws by employing: warning letters, application notices, official cautions and prosecution. The enforcement practice is furthermore taken for mitigating the pollution effects of activities and preventing additional incidents and pollution that might cause severe environmental damages or harmful impact to human health (Ferrey, 2010).

They include practices like: Prohibition notices for stopping or preventing illegal and pollution activities; project (antipollution) notices for enforcing activities required for mitigating or preventing severe environmental damage or diminishing of human health; plus, suspension or cancellation of a permission. Execution activities should be proportional to the overall potential or real destruction to the environment, with regards to the levels of noncompliance or offenses that have been allowed. The standards by which UK regulators make enforcement actions must be law compliant to avoid any violations.

Both MCERTS and OMA Environment laws and regulations in the UK are determined by the Cabinet Agency Enforcement Concordat. The Particular supervisors' policies that should be followed include: the Environmental Agency's Enforcement and trial policy (EA 1998): SEPA's protocol statement on enforcements (SEPA 2002). It must further be mentioned that in Scotland, environment cases are usually sent to Procurator Fiscal department for consideration. For official prosecution, the magistrate's court imposes a maximum penalty of £20,000 or 3 months' prison sentence. For the Crown Court, the penalties are even harsher with an unlimited penalty and probable two-year prison sentencing (Ferrey, 2010).

1.4.3. Protocols of Modern Regulation

The UK government's Better Regulation Taskforce (BRFT, 2005) operates an independent body whose mandate is advising the Government, especially on activities ensuring that regulation and enforcement are done according to the five Elements of Good Regulation. As of now, the Better Regulation Taskforce has been replaced by the Better Regulation Commission since 2006 (Atapattu, 2007).

The edicts applied to environmental management are: (i) Proportional (or risk-based) regulators apportion resources based on the risks presented and the measure of outcomes that may be attained; (ii) Targeted (or result-based): regulators utilize environmental measurements to plan and evaluate performance; (iii) Consistent: regulators use the same technique within and between segments, and over time; (iv) Transparent: regulators have predetermined rules and procedures that are clear and accessible to organizations and regional communities; and (v) Accountable; regulators clarify their decisions and outcomes. Supervisors have or are creating risk-based techniques that employ a variety of unique regulation tools (e.g., authorizing, trading schemes, voluntary contracts, education, and advice, plus environmental levies).

Currently, the Environmental Agency (EA) and SEPA correspondingly have established their methodologies for upgrading regulation in 'Provision for the Environment, which is a 21st century model to regulation' plus an 'Objective for Regulation: Safeguarding and Improving the Ecosystem through Regulation.'

In the modern technique for regulation, operators take full accountability for their environmental output, showing to supervisors that they can efficiently manage environmental factors, utilize resources effectively, and work in a socially viable way. Regulators measure the performance of structured organizations and provide incentives and sanctions comparable to the risk presented to the environment plus observed principles of performance (Sancin, 2012).

The Environmental Agency (EA) methodology for compliance assessment is reliant on the opinion that effort, as well as resources, normally are directed towards risks and environmental impacts. The goal is to promote and reward improved environmental performances and results; making good use of compliance review tools; simplifying regulation measures for companies and supervisory staff; as well as targeting efforts at actions that present the highest risks to environmental welfare.

The EA OPRA Schemes Operator Performance, as well as Risk Assessment (OPRA), have progressed from models formulated distinctly for Integral Pollutants Control (IPC) and Waste Managing Licensure (OPRA for Refuse). The scheme evaluates both environmental risks of an operation, plus the operator's performance through data from operators, reviewed by the (EA) Environmental Agency.

Furthermore, OPRA evaluations create a risk profile, which is founded on these particular five attributes; (a) complexity: the complexness of activities and procedures done and regulated; (b) discharges: allowed for all environmental media including air, land, and water; (c) locality; Closeness to and environmental receivers sensitivity; (d) operator output: the magnitude and efficiency of the operative's environmental management structure-encompassing compliance history, operations, and maintenance, competency, and training, crisis planning, performance appraisal and external reporting; also e)compliance rating: complying with permit stipulations, compelled by the figure as well as significance of license breaches, categorized through the Common Compliance Scheme (CCS) (Sancin, 2012).

Every attribute is scored for producing a "band" range (A to (e) whereby 'A' equals low risk and great performance, while 'E' is great risk and reduced performance. The marks for every attribute are factored to produce the general OPRA risk outline (A to E). Data from OPRA evaluations are utilized in two unique ways: (i) planning compliance evaluations; and (ii) Determining application and yearly subsistence rates-using the model that reduced risk needs lowered Agency regulatory efforts and, therefore, lesser fees are levied on the operator.

Direction on the protocols and usage of OPRA is offered on the (EA) Environmental Agency's online page at www.environment-agency.gov.uk/epopra, there's also some association between the OPRA and environmental managing systems. Besides, the rate of credit provided by the (EA) Environmental Agency reveals the minimal effort in defining permit applications, because of the information produced by EMS, including the probability for minimal compliance evaluation effort (Ferrey, 2010).

The Credits presented to operators with an EMS ratio from: a low range credit for some 'in house' structure; to much higher credits for ISO 14001:2004 licensed by IAF accreditation certification body; plus, up to a full credit for the EMAS registration system. It's meant for extending credit going into the future under OPRA, taking note of EMSs founded on BS 8555 which are examined within the IEMA Acorn Schemes and industry Practice Codes agreed with or identified by the (EA) Environmental Agency.

The process can be done following the re-assessment of Operators Performance (administration systems) attribute after the results of the REMAS program. REMAS generally is a 3-year European program to give evidence that connects environmental management structures and environmental performances. Moreover, MAC, CAPs, and CCS complying assessment procedures (CAPs) have been identified as work planning tools that allow for compliance assessment efforts to be categorized based on risk, ecosystem outcomes and location needs. Besides, CAPs allow for targeting of complying assessment effort, as well as resources, depending on factors such as: (i) local issues and goals (ii) OPRA risk profiles; and (iii) priority outcomes (ecological and regulatory) for manufacturing sectors.

The procedure for assessing compliance, meanwhile refers to a reliable technique for evaluating compliance with licenses through an OPRA profile, plus related directly applicable laws. The methodologies would be applied originally for PPC as well as Waste Management Licensure, alongside CAPs. Compliance classification scheme or (CCS) categories and records cases of noncompliance taking note of permit and license stipulations by operators into a total of five categories. Basically, the scheme offers a consistent context for implementation of EA' implementation and prosecuting policies. It further offers a reliable link to the complying rating attribute in OPRA scheme.

SEPA OPRA is also similar to Environmental Agency's Scheme. The dual modules of SEPA'S scheme include, Operator Performance Appraisal (OPA) and Pollution Hazard Assessment (PHA), roughly equal to the protocol of 'hazard' attributes plus the operator performance qualities of the Environmental Agency Scheme (EAS) (Louka, 2006).

SEPA utilizes OPRA to allot its inspection resources depending on a blend of intrinsic risks of the location and operator's performance for management of these risks. Generally, assessment frequencies are limited for operators with positive OPRA ratings. SEPA doesn't utilize OPRA for other reasons like the preparation of licenses. In respect to EMS and OPRA, SEPA initially had to await the results from REMAS program before deciding whether credit is to be provided under OPRA, particularly for operators having a well-recognized EMS, SEPA considers that EMSs, especially those approved or verified, may play a key role in producing gradual progresses in the environmental output of organizations.

Nevertheless, a direct linkage between possession of EMS and enhanced legal compliance still must be established. In spite of this, SEPA considers that operators with an EMS may likely receive enhanced ratings under the OPA aspect for criteria like management and training. There's hence some ratio of credit established into the existent system. SEPA's OPRA procedure is captured in the 'Environmental Risk Appraisal Manual,' which may be seen on the SEPA official site (Louka, 2006).

1.4.4. Local Authority Air Pollutant Control (LAAPC)

The risk assessment process is normally applied by local administrations in Wales and England for determining the relative amount of risk related with procedures that are structured under the Local Atmospheric Pollution Control Regime. Additionally, the technique assigns some amount of proposed 'regulation effort' to personal processes (high, moderate, or low) based on their comparative risks (Fisher, 2017).

The technique relates to efforts applied in regulation of procedures upon being authorized-meaning what's captured by the subsistence aspect of Local Authority Atmosphere Pollution Control (LAAPC) rates and charges. Besides, risk appraisal applying this technique is reliant on the overall nature of the procedure, and how it's managed; this is subdivided into other key aspects. The first is Environmental Impact Appraisal or (EIA) that's concerned with the prospective environmental effects of the process based on type, amount of upgrading for meeting regulation needs and location; plus Operator Performance Appraising or (OPA) that relates to how expertly the operative is able to manage possible environmental effects of the process (Fisher, 2017).

1.5. LAWS THAT ESTABLISH COMPLIANCE 3

1.5.1. State Statutes and Regulations Implementing the Federal Statutes 3

Figure 1.3: Nature is interconnected and must be handled with care.

Source: https://www.baurome.com/environmental-law-course.

The U.S. environmental law is concerned with legal standards for protecting human health and improvement of the American natural environment. Though prone to criticism locally and oversees on matters of protection, execution, and extra regulation, still the country remains to be a significant contributor of environmental law expertise and experience. The U.S. Congress has ratified federal statutes meant for addressing pollution management and remediation, such as for instance, the Clean Atmosphere Act addressing (air-pollution), Clean Water Policy for (water-pollution), as well as the Comprehensive Environment Response, Recompense, and Liability Act (Fisher, 2017) (Figure 1.3).

Furthermore, there are federal policies that govern natural resources usage and bio-diversity which are highly influenced by environmental protocols, such as the Endangered Wildlife Act, National Forests Managing Act, and Coastline Zone Management Act. Additionally, the National Environments Policy Act (NEPA), determining EIA in actions done or approved by American federal government, might implicate all of these aspects (Fisher, 2017).

Federalism in America has played a key role in shaping national environmental statutes. Many federal environmental legislations use cooperative federalism systems-a number of federal regulatory protocols are managed in conjunction with American states. Moreover, the states largely have passed their unique laws covering aspects not anticipated by federal law. It covers aspects where Congress had probably acted in restricted way (e.g., state site cleanup legislation for handling sites outside Superfund), plus where Congress may have left regulation mostly to the states (e.g., water resources law).

Brief History: The history of environments legislation in America can be traced back to its original source in common law policies, for instance, the decree of nuisance and public trust policy. The original environmental statute to be enacted was the Rivers and Harbors Enactment of 1899, a system that was essentially succeeded by the Clean Waters Act. Nevertheless, most current primary environmental statues, like the federal statues mentioned here, were passed during the period spanning from the late 1960s up to the early 80s. Before the passing of these statues, many federal environmental laws weren't almost as all-inclusive (Bodansky, 2010).

Additionally, Silent Spring, which is a 1962 book written by Rachel Carson has frequently been noted for launching the environmental campaign in the U.S. This volume documented the negative impacts of pesticides, particularly DDT, on the birds and nature in general. Among the most impactful environmental tragedies of the 60s was Santa Barbara oil spillage which occurred in 1969, causing considerable public reaction as Congress was deliberating various aspects of environmental legislature.

One particular litigation that has been extensively identified as among the first environmental examples is the Scenic Hudson Protection Conference vs. Federal Power Contract, mentioned in the 1965 2nd Circuit Court of Pleas, before passage of the main federal environmental decrees. This case helped stop the building of power facilities on Storm King Mount in New York State. It has been defined as inspiring environmental litigation and facilitating the creation of legal principles of repute for delivering environmental claims (Bodansky, 2010).

Besides, the Scenic Hudson lawsuit is believed to have aided in inspiring the enactment of the National Environments Policy Act or (NEPA), including the development of environmental advocacy bodies as the Natural Reserves Defense Council. Nowadays, legislation from almost every layer of policymaking in the U.S. relates to environmental matters.

In addition, Congress has passed various landmark environmental regulation programs, however, there a number of other federal policies which are equally significant, despite being less comprehensive. Simultaneously, the legislations of up to 50 states have ratified countless comparable sets of legislation. These statewide and federal policies normally are foliated with level upon level of administrative regulation. In the meantime, the U.S judiciary system evaluates not just the legislation enactments, however also the managerial decisions of agencies that deal with environmental matters. It's widely believed that where typical statutes and regulations terminate, common law starts (Bodansky, 2010).

Federal environmental law statutes: Through federal statutes, US federal agencies impose protocols in the Policy of Federal Regulations which fill out the wide programs ratified by Congress. Key among these aspects is Subject 40) of the Federal Regulations Code, consisting of the (EPA) EPA regulations. Other key CFR sections are Title 10 touching on energy, Title 18 (Preservation of Energy and Water Assets), Title 21 (Foods and Medicines), Title 33 (Water Routes), Title 36 (Woodlands, Parks, and Public Properties), Title 43 (Communal Lands: Interior), plus Title 50 (Fauna, Flora, and Fisheries) (Fisher, 2017).

Both federal and state courts have played crucial roles in formation of environmental laws in the U.S., in a number of cases, they have been ideal at resolving substantial controversy concerning the submission of federal environmental rules in favor of ecological interests. The resolutions of the Supreme Court in matters like Calvert Cliffs Coordination Committee versus. U.S Atomic Power Commission (widely reading the procedure needs of NEPA), including Tennessee Valley Authority vs. Hill (widely reading the Threatened Species Act), plus, much more lately, Massachusetts vs. EPA (demanding EPA to reconsider greenhouse gases (GHGs) regulation within the Clean Atmosphere Act) have all had significant policy impacts way beyond the evidences of the specific case (Fisher, 2017).

Common law: The general law of torts serves as an essential implement for resolving environmental disputes falling beyond the margins of regulated activities. Before contemporary explosion of environment regulations, the principles of nuisance (private or public), trespassing, negligence, and severe liability presented injury and allotted responsibility for actions that nowadays may be regarded as pollution, and probably determined by regulation regimes (Fisher, 2017).

Majority of these doctrines still remain applicable, and more lately have been applied by plaintiffs who seek to enforce liability for outcomes of global climate variations. The common law furthermore continues playing a major role in U.S. water legislation, specifically in the context of riparian rights as well as prior appropriation.

In the U.S., responsibilities for administration of environments legislations are subdivided between different federal and state bodies with variable, overlapping, and occasionally contradictory missions. EPA is considered the most renowned federal agency, having jurisdiction over most of the country's nationwide air, water, residue, and toxic substance programs (Louka, 2006).

Alternative federal agencies, like the American Fisheries and Wildlife Services and National Parks Service follow primarily conservational goals, while still others, like the U.S. Forest Service and Lands Management Bureau, seem to concentrate more on valuable usage of natural resources. Additionally, federal agencies work within the bounds of federal jurisdiction. Case in point, EPA's jurisdiction within the Clean Waters Act is generally restricted to "waters in the U.S." For many cases, federal laws permit for stricter regulation by states, plus of transferring certain federally mandated roles from federal administration to state management.

The American state governments, thus, managing state law approved under state police influence or federal laws through delegation, uniformly cover environmental agencies. Besides, the magnitude to which state environment laws are reliant on, or vary from federal law may differ from one jurisdiction to another (Louka, 2006).

Therefore, while a permit for filling non-federal wetlands may need a license from one state agency, bigger, and more complex measures—for instance, the building of coal-based power plants—may need approvals from various federal and state bodies. As for enforcement, in America, environmental law violations are normally civil offenses, which result in cash penalties and, possibly, civil sanctions like injunction.

A number of environmental laws furthermore provide for criminal charges for severe violations. Some federal decrees, like the Clean Waters Act, furthermore allow U.S. citizens to file lawsuits against violators, if the administration has failed to implement enforcement guidelines. Environmental bodies normally include separate enactment agencies, with duties like assessing acceptable activities, doing compliance inspections, providing citations and prosecution of wrongdoing (criminally or civilly, based on the violation). The EPA's Bureau of Execution and Compliance Standards is one example of this service (Bosselmann, 2016).

Others, like the U.S Park Police, perform more conventional law enforcement actions. Mediatory processes for environmental violations furthermore are typically addressed by the organizations themselves, under the pretext of administrative legislation. For some instances, appeals may further be addressed internally (for instance, EPA's Environment Appeals Board). In general, final agency determinations might consequently be appealed with the proper court.

1.5.2. Torts 4

Definition: Environmental torts can be described as damages to property or property value, they are related to toxic torts which means injuries to people. Even though linked by causation, they depend on different causation analysis and damages. Environmental torts have the ability to cause either direct or indirect destruction to property values.

Furthermore, pollutants may be visible discharged towards a particular property, like oil or chemical spillage. Even though a property might have been cleaned, plus declared non-hazardous, still there could be enduring stigma impacts to property value. Purchaser fear may diminish a property's market rate even when those buyer threats are unfounded.

Present operations, previous discharges and previous contamination of industrial spots may damage the value of adjacent properties even when there's no instant discharge. The damage can come from existing noise or smells, or possible risk of imminent accidents. Scholars have analyzed the effect of environmental destruction on the overall value of properties which are affected by Superfund spots, landfills, pipeline breakdowns, refinery releases and pesticide plant pollution.

Economists and statisticians apply a variety of tools such as multiple regression examination, and extensive appraisal information to determine the effects of stigma destruction to a property. They can also explore whether a present property owner has in any way been damaged, or whether the current owner bought the property when the environmental factor had already been factored into the buying price.

Toxic Torts: The toxic tort basically is a personal harm case that involves injury triggered by exposure to toxic products like chemicals, pharmaceuticals, toxins or pesticides. Additionally, the injured party may sue for lost incomes and medical costs.

Plaintiff and defense lawyers may further help with professional inquiry in toxic tort instances. They will determine possible loss of wages and rationality of medical costs. Other professionals like life care planners, physician consultants and economists work together to determine the wounded party's physical or cognitive impairments, prospective healthcare needs, and existing value rate of the care.

Context and Structure: A tort, largely speaking, can be considered to be a civil wrong. The description nonetheless leaves a lot to be desired-case in point, one solution for a tort case is an injunction, preventing harm before it happens instead of punishing it. The description also covers breaking of accord-even though contract law works as is its own unique field. Notwithstanding these shortcomings, the meaning is workable.

Torts are different from crimes that are normally regarded as a wrong committed against the society at large, and prosecuted through state power. Furthermore, torts are prosecuted solely by the injured individual (or their legal representative), where the damages sought would be fiscal and compensatory, instead of imprisonment. The torts and felonies may be prosecuted simultaneously, yet not mutually, plus civil recovery isn't precluded by any effective criminal prosecution.

Considering the risk of imprisonment, torts have another risk of punitive injuries - most merely described as the recovery which doesn't emanate from actual damage however rather is meant for deterring future tortious activities. Most of the current debate around tort reform is based on the concept of punitive damages: or how much can one person be able to profit in order to prevent another individual's or entity's unclear future conducts?

The decision on if punitive damages must be capped (more than the limits established by the Supreme Court within constitutional due procedure) doesn't alter the fact torts play a key role in aiding to define the standards of acceptable conduct in society. The function of torts in influencing law is another basis of dispute amongst tort reforming camps, however the reality of torts law for todays; legal setting is undeniable.

Torts can be subdivided into three categories which are Intentional Torts, Negligence, and Strict Liability. Simply put, negligence is behavior a person is held liable for even though there's no harm that was intended, besides intentional torts cover determining behavior that causes liability, though strict liability doesn't cover fault or a practical measure of care.

1.5.3. Laws That Enforce Compliance (Permits, Prohibitions, and Penalties) 6

Enactment of environmental law continues to be a major challenge, particularly in developing countries because of factors like lack of institutional capability, minimal competence on the part of enforcement officers to enforce legislation, plus lack of information or national guidance materials concerning execution.

All these work to decline the efficacy of the laws around protection and management of the environment, while environmental degradation becomes the final result. Enforcement thus is a main factor for allowing national environmental legislation to achieve their goals of managing or safeguarding the environment from possible deprivation.

There are three main thematic subjects which are: joint enforcement activity; legal factors of environmental destruction calculation and compensation; plus, measuring the efficiency of enforcement activity. These positive practices were first produced at the Africa-Asia cross-regional meeting of professionals on enactment of environmental law, which was conducted in 2015.

Throughout the previous years, to enhance the application of, agreement with and enforcing of environmental laws, a number of countries have implemented and practiced different integrated enforcement tactics such as integrated permitting, mutual inspection and enforcement, involvement of the public in implementation, promotion, and rewarding positive compliance, including collaboration with private players.

Measuring the compensation for environmental destruction remains to be a tough matter for judicial structures in their everyday adjudications. In an endeavor to regulate and control the practice, some nations have used legislative tools to direct judges and enforcement officers in determining and compensating environmental destruction.

They endorse various laws on environmental obligations and reimbursement of environmental damage, creating the legal structures for calculation and compensating of environmental damage, plus develop the procedures defining environmental damage. Some other aspects where countries have embraced regulatory rules and established guidelines are: who carries responsibility to reimburse, compensable costs, compensation entitlements, procedures, and who performs evaluation and methodologies.

There's always an application gap between legislation goals and efficiency of environmental policies. What's effectiveness of application and enforcement activities remains to be a test for lawmakers and implementation agencies to answer. For addressing this concern, some nations signed to the (OECD) Organization for Economic Cooperation and Development have established and used performance measuring indicators for measuring environmental enforcement.

Nonetheless, countries in Asia and Africa have restricted practices in formally calculating the success of enactment actions. This volume mentions selected ideal practices and delivers them in three main categories which are: (a) inventive joint enforcement activity, (b) legal aspects of measuring and compensating of environmental damage, and (c) measurement of enforcement actions, with particular intention of offering other countries diverse perspectives and options which may be applied to strengthen environmental enforcement systems. By the end of every set of good practices, challenges associated to that matter are identified, plus the recommendations are proposed to address those difficulties.

Joint Enforcement Action: Currently, it should be identified that environmental concerns are more complex compared to the past, plus more nuanced solutions would be needed. Conventional regulatory measures and techniques, in specific command-and-control regulation approaches, still remain to be the most broadly used legal procedure for environment quality management plus pollution control. Amongst the most extensively used conventional techniques for environmental management are compulsory requirements for fulfilling environmental measures, the structure of authorization by licenses, certification or licenses provided by government bodies.

These are known as command-and-control processes. In reaction to variations in socioeconomic and environmental factors, nations reformulate laws to meet the trials they're facing. There must be improved scientific evaluations of the ecological connections among various environmental aspects/media. Responding to these needs, ecological impact assessment including the application of abatement procedures to diminish environmental pollution and effects also become a key mechanism for pollution management or reduction, plus protecting the environment.

Lately, various new techniques and mechanisms have been identified to complement the conventional regulatory techniques seeking to enhance the application, compliance, and enforcing of environmental legislation. These comprise of: (a) Economic motivations or disincentives for internalizing ecological externalities (b) Coordination among appropriate governmental bodies with shared responsibility through a top-level advisory committee (c) Public consultation systems - done through a variety of mechanisms, like public involvement and engaging in environment decision-making, communal hearing on environmental evaluation of key projects, ad hoc conferences or web-based social media, up to informing governmental decision-making.

1.5.4. General Purpose Criminal Laws 6

Definition: To value the progress of criminal law matters as a tool of environmental safety, it's important to start with some short historical context. Even some casual glimpse at legal history divulges a conventional resort to criminal law, being a primary and effectual way of solving different economic, social, and political matters.

The tendency could be a result of 18th century "absolutist" legislative thinking, based on which it remained the duty of the independent to actively encourage the common good, therefore regulating controlling various features of civil existence. Under the absolutist model, those who violated the rules recommended for the positive order of the society were chastised, incarcerated, or fined.

Inevitably, in a political structure where criminal rulings are subject to the whims of a Leader, who also holds judicial and legislative influence, many transgressions were regarded as criminal offenses. It wasn't until many constitutions in continental Europe actively incorporated the rule of power separation, enacting standard and contemporary legislation, that it became feasible to develop a convincing and dependable distinction between criminality law and other aspects of the legal arrangement.

As criminal law grew to be seen as the "ultimate answer," still it was put under exclusive court control, and every possible violation of social standards meriting punishment were combined in one criminal code. Consequently, a number of socioeconomic, and, certainly, environmental legislation created a number of corresponding penal standards outside the typical criminal codes.

With time, these penal standards were enacted to safeguard the environment in states like Spain and France. Over the recent times, rapid industrial growth has prompted safeguarding of the environment as a factor of immediacy in most European states. A number of these nations have embraced criminal law in times other procedures proved to be insufficient or ineffective.

The tendency is captured in this excerpt from the 10th International Assembly on Comparative Law. It mentioned "When people encounter an emergent disastrous situation, they are normally shocked and are quick to rush to punitive legislation to overturn it. In fact, people seem to believe that the issue is solved when they have a fresh penal law." It was the case when the European Council approved Resolution (77) of Chapter 28, developed to promote "ultimate imminent harmonization of statutes of the member countries of the European Council in this sector. "

Given the significance of environmental law, the European Commission considers criminal law to be an effective technique for safeguarding the environment. Besides, the penal security of the environment turned to be an issue of actual concern in Europe during the year 1978, when Commission of Ministers from the European Council adopted Clause (77) 88 on the impact of criminal law in protecting the environment.

Key developments: The Commission of Ministers first reflected on the need for protection of human health, including animals and plants, plus the splendor of landscapes. Secondly, the Committee observed that different facets of present-day life, particularly industrial development, involve some degree of pollution that's especially hazardous to the community.

Furthermore, the Commission resolved that the need to adopt criminal legislation as an "ultima quotient" existed where other processes are ignored, ineffectual, or incorrect. Depending on these resolutions, the Ministers came up with these particular endorsements to member states administrations:

- Investigation of criminal fines for destruction to the environment, plus whilst maintaining the conventional penalties of fine including imprisonment (probably conditional) in severe cases: (a) Introduction special kinds of pecuniary penalties, like daily fines (astringes), suspended fines, and conditional fines; (b) allocating proceeds from pecuniary fines for pollution to ecological uses; (c) introducing measures like restoration of previous status possibly arranged in conjunction with a deferred custodial fine, work for the advantage of the community, prohibitions (as principal penalties), plus publication of opinions; (Organisation for Economic Cooperation and Development, 2009) Available at: https://www.oecd.org/env/outreach/42356640.pdf

- Re-examining principles of criminal obligation, with a perspective, in essence, to the likely introduction in some cases of the problem of corporate systems, private or public;
- Examining of advisability of criminalizing activities and omissions that culpably (purposely or negligently) put the existence or health of people or property of considerable value at considerable risk;
- Re-examining criminal processes in aspects" of environmental safety and particularly: (a) creating specialist court branches and public prosecution offices for dealing with environmental matters; criminal charges for safeguarding the community's interests; (c) developing a special criminal record of individuals who have been convicted for causing pollution, unique of the overall criminal catalog; and (d) omission from amnesty of severe environmental felonies.

Resolution (77) of Chapter 28 of the European Commission's resolutions can be regarded as the beginning point for a highly systematic application of criminality law for protecting the ecosystem in Europe. At the very least, it directs attention to the benefits which some member states can derive from steadily compiling in one collection, with particular emphasis on criminal provisions related to environmental safety with a perspective to. An ultimate future harmonization of statutes of member parties of the European Council in this sector (Bosselmann, 2016).

The significance of using criminal laws for protecting the environment can be evaluated by considering various trends that co-exist in contemporary criminal law. It's been proposed that there subsists in contemporary criminal law two opposing trends "which seem to conflict however are, actually, linked by a single philosophy; conversely, there's a tendency for decriminalizing certain offenses, especially those considered conventionally as comprising no victims (for instance, vagrancy, personal use of some drugs). This activity is called depenalization. The procedure of depenalization replies to two basic requirements.

One is substantive, consisting of reconciling the welfares protected by penal authorizations in the framework of what are considered to be prevalent primary values. The alternative is practical, or praxeological, consisting of reduced caseloads such that courts can focus on more severe crimes. Both of requirements are fully consistent with the secondary trend consisting of treatment "as criminal many kinds of behavior associated with technological actions which may severely impair the health, security, and welfare of the public (Louka, 2006).

Environmental protection by way of penal measures is a prime illustration of this projection of criminal law." Regardless of the application of criminal sanctions towards safeguarding the environment, criminal penalties only complement other regulation measures. In fact, criminal felonies in the environmental perspective must "promote and supplement existent regulatory offenses that relate principally to water, air, and land pollution, plus nature preservation."

Similarly, experts mentions that "penal law contributes just a limited role plus even if it contributes, it does this mainly in conjunction with administrative guidelines." The secondary function of criminal law can be demonstrated by the way that normally, only the most severe cases are directed to criminal courts for deliberation (Louka, 2006).

1.5.5. Enforcement Options 6

Figure 1.4: Environmental law practices may vary from one country to another.

Source: https://www.istockphoto.com/photos/environmental-law.

In order for environmental legislation to 'work,' it shouldn't just be well developed but also efficiently implemented and enacted. For regulators, the latter involves two key tasks. First, there's the inquiry of where it's ideal to apportion resources and the sectors and businesses to target for assessment—often seen in regards of 'how they can get the greatest bang for their regulatory buck' (Louka, 2006) (Figure 1.4).

Secondly, strategies should be designed as to how inspectors can go about the work of intervening in matters of regulated establishments. Should they for instance, seek strict enforcement or discuss outcomes through counsel and persuasion? Briefly, the two key setbacks for regulators are deciding where to intervene, plus how to intercede in the matters of controlled organizations.

As concerns the first challenge, there's now something approaching agreement that the most ideal way to apportion scarce regulation resources is by risk-based assessment. This involves 'developing decision-making structures and procedures for prioritizing regulatory activities, plus deploying resources, primarily related to inspection and implementation, based on the evaluation of risks which regulated organizations present to the regulator's goals' (Louka, 2006).

In the UK, risk-based guidelines were endorsed by Hampton Review, then afterwards encapsulated in the statutory practice code. Nevertheless, this technique is by no way restricted to the UK, plus a growing ratio of OECD nations are seeking to create coherent structures for the management of risk in regulation policy.

Furthermore, risk-based regulation contains its own limitations and strengths, including complexity of the risk evaluation process and grade of information accessible for basing it on. Nevertheless, many regulators and policy developers, at least in European countries, would consider it as estimating best practice in regards to where to apportion majority of their resources (Bodansky, 2010).

Contrarily, the second role, deciding how to intercede in the matters of regulated organizations for ensuring compliance and expediting enforcement (intervention strategy), has further been provided far less consideration, at least by supervisors themselves.

Besides, there's nothing distantly approaching consensus in regards to how it must ideally be determined. On first sight, this may be unforeseen, since intervention procedures will have a vital bearing on regulation results.

This comparative neglect by regulators (shown in minimal sophistication or consistency of such procedures in many places) is possibly even more surprising, considering that there's a rich scholarly literature that has discussed this matter for several decades (Bodansky, 2010).

Furthermore, individual supervisors 'on the ground' might not really conform to their agency's specified intervention strategy, plus therefore, a related question involves how inspectors act in practice (including why) and exactly how to restrict inspectoral preference. How, for instance, can preference be contained without enacting mechanistic rules and motivating inspectors to 'abide by the book?' In particular, this inquiry has been a matter of substantial research, even though it hasn't yet formulated anything approaching consensus. (CompassPoint Nonprofit Services, 2012) Accessible at: http://www.gnof.org/wp-content/uploads/2015/06/Supervisor-Roles-and-Responsibilities-Packet.pdf

With regards to methodology, intervention strategies are necessarily established at regulator levels and normally established in a formal strategy, which is addressed for field officers and typically also to structured organizations as well as other stakeholders. Therefore, any relevant policy is normally to be discovered in publicly accessible documentation given by environmental bureaus themselves (Sancin, 2012).

Consequently, the first call port in investigation of intervention strategy is formal documentation, through websites, press releases, yearly reports and other periodicals, supplemented, where accessible, by public discussions of agency policy with senior staff. Nevertheless, while in the example of agencies such credentials visibly articulate a specific intervention policy, this isn't always the matter. Where no policy is clearly pronounced then external sources, such as scholarly research, commentary by previous agency officials plus by regulated bodies or their associations, may also be used to get improved comprehension of intervention policy.

2

Principles of Environmental Law

CONTENTS

2.1. INTRODUCTION

Environmental law is a term that includes all aspects of law that protect the environment. Environmental law is shaped by principles which focus on the management of resources that are natural and exist within the environment some of which include forests, minerals, wildlife, and fisheries. Aspects such as impact assessment are components that are important to environmental law.

Throughout history there has been creation of laws meant to preserve the environment. For example, the common law, the primary protection was found in the law of nuisance which only allowed action to be taken privately for damages in the event there is degradation done to land.

In June 1992, government leaders and representatives from over 100 countries and over 30,000 participants met in Rio de Janeiro at a United Nations conference on the environment and development also referred as to the Earth Summit. At this conference there was recognition of the need combine or bring together both economic development and protection of the environment in order to achieve sustainable development.

The Earth Summit also recognized the importance of international environmental law as a means that would help foster and achieve development that was sustainable. Public international law comes from different sources for example conventions done at an international level, laws that are customary, general principles of law acknowledged by countries that have achieved civilization and law-related judgments or the work of qualified publicists. The above categories have been important in the international environmental law development. Some of the laws include more than 800 agreements that are either binding between two countries or between multiple countries which provides guidance to address different issues related to the environment. The issues range from sub-regional, regional, and also global issues. Customary international law has led to the emergence of different environmental principles and concepts. Environmental law is still a growing field and international laws related to the environment still continue to develop from less traditional sources which are less binding. For example, resolutions, and declarations that are created and issued by international organizations such as the United Nations environmental program, the world health organization and other agencies that are international. These agencies have played a significant role in the creation of environmental principles of law internationally but most of the time they are not binding legally. When the laws are repeated, practiced, incorporating them to local legal systems, the said principles may be brought forward as customary law.

There are different aspects related to the environment they include, environmental impact assessment (EIA), water resources, mineral resources, forest resources, wildlife, fish, and game. EIA assesses the consequences that results from human interactions with the environment. It assesses plans, policies, programs or actual projects before a decision is made to move forward with an action that is proposed. The consequences can either be positive or negative. It is a tool used in the management of the environment and forms a part of approving projects and making decisions. It is sometimes governed by rules of administrative procedures with regard to participation by the public and documentation of decisions which can be made subject to review judicially (Atapattu, 2007).

Law related to resources such as water direct the use and ownership of water resources that is both surface and groundwater. Regulations created may include areas such as conservation of water, restrictions, and ownership regimes. Laws governing activities related to forests guide the management of forests and also the use of forest resources. The acquisition of forest land and burn practices that are prescribed are guided by ancillary laws. Government agencies bear the responsibility of planning and implementing forest laws in relation to public forest land and therefore they maybe be involved in the inventory of forests, planning, conservation, and oversights of the sale of timber. Initiatives are also taken to prevent or reserve deforestation. Environmental law can also govern the impact human activities or interactions on wildlife and plants for example the degradation of wildlife habitat degradation. The laws also offer protection to plant species. They can also be enacted to protect biological species from extinction. These efforts also entail the awarding of special conservation status to species that are being protected, the laws will lead to creation of centers of wildlife and plant conservation reserves. It also supports and enhances the protection of wildlife and plants and development of laws which prohibit the poaching and trafficking of wildlife. Environmental laws also regulate the right to hunt down and kill of fish and wild animals which are a source of game meat. This type of laws restricts the days of harvesting fish and hunting wildlife, the number of species caught per individual, or the tools, equipment used especially for fishing activities. They provide a legal structure for giving out licenses and fees for those involved in fishing and hunting of wildlife which in return is used to fund efforts for conserving resources in the environment as well as the process of obtaining information on harvesting which are employed in the management of wildlife (Atapattu, 2007).

Environmental law came up as a result of the emerging awareness and concern over issues that globally impact the environment. Environmental laws have been development overtime and with relation to issues brought forward and changes in environmental systems and for this reason efforts have gone into the identifying and developing key concepts and principles that show similarity to environmental law as a field. There are a number of environmental law principles which are not always recognized and accepted universally but are important in understanding the field of environmental law globally. The group of principles and concepts have been used to shape environmental law overtime are outlined in publications such as our common future-dated 1987 and published by the world commission on environment and development and also the Rio earth summit declaration of 1992 (Atapattu, 2007).

2.2. THE PRECAUTIONARY PRINCIPLE

2.2.1. Description and Elements of the Principle

The precautionary principle of environmental law was formulated in the Rio declaration and it is one of the most frequently encountered and controversial principles. The precautionary principles state that: "in order to offer protection to the environment, an approach of precaution shall be widely applied according to the ability of a country. If there any serious threats that are existent and can cause damage cannot be reversed, lack of full certainty related to science shall not be used as a reason delaying measures that are cost-effective to deal with destruction of the environment." The principle uses the better safe than sorry approach contrary to the wait and see approach to protecting the environment. This principle plays a role in any debate over the need for regulating the environment (Sands, 2003).

The precautionary principle plays a role which is significant in determining whether the process of achieving sustainable development is sustainable or not. Its formulation was seen as a response to the growing appreciation of the scientific uncertainties related to degradation of the environment. The principle reflects two insights that are important the first is the reliance certainty related to science as a means of determining measures of response and two the irreversible impacts that are consequences of not taking preventive actions at an early stage.

It underlies sustainable development which necessitates the stopping and prevention of developmental activities in the event that they lead to environmental damage which is either serious or irreversible. The creation of the precautionary principle marked a shift in the international environmental jurisprudence. The precautionary principle was shifted from the principle of assimilative capacity (Sands, 2003).

The precautionary principle can be applied based on a criterion that is as follows:

- Risk Management: The principle should be invoked as part of analysis of general risk that leads to making a decision to favor or be against a decision related to the environment.
- Scientific Assessment: Taking precaution should be based on assessment that is scientific and that has been done comprehensively. Scientific assessment should be done to the furthest scientific extent possible.
- Proportionality: measures for managing risks should be proportional to the protection level selected or chosen. Other factors taken into account include the cost, effectiveness, and duration. Precaution measures may include providing additional safety especially for marginal groups.
- Coherence: The measures taken should be consistent with those adopted in situations that are similar.
- Trade Restrictions: Precaution should be invoked if there exists potential risks. This should be used to justify decisions that are arbitrary and therefore it should not be disguised as interests related to trade.
- Intelligibility and Review: Measures that respond to environmental problems should be founded on facts and intelligible. The measures should be regularly reviewed in the event of scientific findings that are new and modified when necessary.
- Transparency: There should be transparency, participation, and accountability in the analysis of risks and its management.

2.2.2. Origin and Application of the Precautionary Principle

The assimilative principle was stressed in legal actions at early times for protection of the environment. A conference was held by the United Nations in 1972 at Stockholm. The conference resulted in to the adoption of the

Stockholm declaration which had 26 principles. The Stockholm declaration contained the principle of assimilation which was the 6th principle. This principle speculated that science would provide those making laws with the information that is important and the instruments needed to avoid over exploitation of the environment beyond its capacity to deal with the impacts. It also makes the assumption that technical advice that is relevant would be available if a prediction was made in relation to environmental harm and time would also be available to prevent such harm. The assimilative capacity of the environment was based on the assumption and belief that scientific theories would provide the solutions for restoration of the ecology whenever pollution and environmental degradation occurred if they were adequate and certain. It was also built on the foundation of certainties and adequacies in science but it suffered setbacks due the uncertainties and inadequacies of science that were visible in the environmental context (Beder, 2013).

Shift from the principle of assimilative capacity.

The inability of the assimilative capacity to provide solutions to environmental problems prompted changes to in concepts related to the environment during the Stockholm conference period in 1972 and the Rio conference 1992. The basic shift initially occurred between the year 1972 and 1982 (Beder, 2013).

It is not possible to precisely determine what effects or impact an activity will impact on the quality of the environment or on the health of humans. Similarly, It is also not possible to know generally, for example, whether a precise level of air pollution will result in morbidity or mortality rates related to diseases of the respiratory tract, or if given levels of water pollution will decrease the aquatic population or whether the presence of oil in an area that is environmentally sensitive will affect or significantly disturb native wildlife that inhabits the area. The emphasis shifted to the precautionary principle as the eleventh principle adopted by a majority of 111 votes with 18 abstentions and one vote against it by the United States on the 28th of October in the year 1982. The charter was overly supported by many countries that were still developing at the time. The world charter for nature supported the prevention of activities which had the potential to cause destruction that was irreversible to the environment. The principles ensure that a substance or activities that lead to environmental degradation are prevented even when there is no scientific proof that lines the particular substance or activity to environmental damage. The words substance and activity means things that are introduced or released to the environment as a result of activities by humans or interactions with the environment. The shift in emphasis of precautionary principle was reiterated in the Rio conference in 1992 as the 15th principle (Beder, 2013).

In simple terms the precautionary principle requires that if there exists strong suspicion that certain activities will lead to environmental degradation then the activity should be prevented or control rather than waiting for incontrovertible evidence to do the same.

Precautionary Principles in International Instruments: It appeared internationally in mid-1980s. It received acknowledgment in 1985 at the Vienna conference for ozone layer protection and involved parties acknowledged the measures they had undertaken to prevent ozone layer depletion at national and international levels. Similarly, in 1987, during the Montreal protocol the members agreed to apply measures of precaution or control of emissions at a global level which lead to ozone depletion.

In the North Sea ministerial conference in 1987, the principle was referred to thrice. In the protection from substances that have the possibility of causing damage and in the third conference in 1990 members agreed to continue applying the principles. In 1991, the convention on the ban of importation to Africa and the control of transboundary hazardous waste movement and management. The convection adopted preventive and precautionary approach to problems related to pollution. In 1992 parties to the Helsinki convention on the protection and use of transboundary water courses and international lakes agreed to be guided by the precautionary principle to avoid potential transboundary impact of hazardous waste release. In the Rio summit four instruments signed also referred to the precautionary principle. Agenda 21 also refers to the precautionary principle in different and multiple contexts for example in relation to marine protection. In the Rio earth summit of 1992 a way was paved for converging the precautionary principle and climate change issue in international law. It encouraged the use of precautionary measures and approaches to anticipate, prevent, and minimize the causes of climate change and mitigation of adverse effects (Sadeleer, 2002).

A reference to afore quoted article was provided in the preamble of the 1997 Kyoto protocol which was guided by the third article of the convention. Parties from the UNFCCC had already endorsed the precautionary principle as a norm. Some of the formulations from this conference were contested by environmentalists and political analysts for being weak precautionary formulations.

The precautionary principle in the United States was included in developing conservation of habitat plans needed under the aegis of the endangered species Act. The principle was invoked by the EC in 1989 when the importation of beef was banned because it was fed with hormones, and in the year 2000, the organization fully adopted the principle as a general principle of international law. The principle was also invoked in the year 1999 by Australia and New Zealand while suing Japan for alleged overfishing of the southern bluefish tuna. The principle, therefore, is essential in environmental and human health protection when it implemented in the field of production and energy resource distribution (Sadeleer, 2002).

2.3. THE PREVENTION PRINCIPLE

2.3.1. Description of the Principle of Prevention

The concept of prevention can be considered an overarching aim that gives rise to multiple legal mechanisms, which includes assessment of environmental harm before it happens, licensing or authorization that conditions that are set out for operation and the consequences of violating set conditions, as well as strategies and policies for adoption. The Concept of prevention has several applications such as setting of emission limits or other products or processing standards and also the use of the available techniques that are the best (Koivurova, 2013).

The principle states that have countries have the right to exploit their own resources and similarly have the responsibility to ensure that their activities which should be within their jurisdiction or control do not degrade the environment of neighboring countries or areas beyond. The precautionary principle is closely related to the prevention principle but they differ widely as the prevention principle requires states or countries to take preventive action against known environmental risks to prevent them from materializing. The precautionary principle on the other hand deals with uncertainty or risks that are hypothetical by constraining possible damaging. Activities (Koivurova, 2013).

2.3.2. Origin and Application of the Principle of Prevention

The prevention principle assumes importance that is overriding in every environmental policy that is effective since it allows nations to protect their environment and the neighboring ones at early stages.

Prevention principle is at the Centre of actions that have been proposed in the Agenda 21 chapters on chemicals and management of waste and also the reduction of risks to health that result from pollution of the environment and hazards which is in 6th chapter of the program areas which protects and promotes human health. The three chapters recommend the minimization of risks to both the environment and to the health of humans through application of preventive measures such as the control and distribution of pesticides; expansion of international assessment of chemical risks, establishment of programs for risk reduction, exchange of information with regard to risks and toxicities of substances, strengthening the capabilities of national management of chemicals and preventing the illegal trafficking of chemicals considered as harmful. It also recommends and promotes the prevention of wastes that are hazardous, it is strengthens the institutional capacities in the management of hazardous waste and prevention of trafficking hazardous waste which is considered illegal (Bosselmann, 2016).

The current formulation of the prevention principle was brought forward in 1972 as principle 21 of the Stockholm declaration on the human environment. Most of the environmental laws have been drafted and implemented in response to catastrophes but it has been noted that preventing harm to the environment is much cheaper, easier, and less dangerous to the environment than taking action in relation to harm that has already happened. The prevention principle is the basis behind the creation of environmental laws governing the generation, transportation, treatment, storage, and disposal of wastes that are hazardous. It also regulates laws on use of pesticides. The convention also contains measures that have a preventive nature such as bans, restrictions, and technology standards to reduce and eliminate persistent substances that are organic from production that is intentional as well as that which is not intentional that are as a result of human activities such as incineration. Participants of the convention recognized persistent organic pollutants as only a fraction of the chemical that are a problem and also agreed to a science-based process of adding other chemicals to prevent further harm that this substance may cause. The use of prevention is one of the best ways to mitigate environmental problems since most human-produced substances that are released to the environment can remain or exist in the environment for many years or a given period of time (Atapattu, 2007).

In 1989 the principle of prevention was the foundation of the Basel convention where one of the subjects was transboundary movement control of wastes that are hazardous. They sort to minimize hazardous waste production and at the same time deal with illegal dumping. It was also an important element which was adopted in the European convention in the third environmental action program in 1983.

The precautionary reasoning is also adopted and applied by measures agreed on from the Montreal protocol on depletion of the ozone layer by substances, this was to help to prevent the ozone layer from being damaged further by phasing out some of the ozone-depleting substances and the framework convention on climate change referred to as the Kyoto protocol which had the main aim of preventing anthropogenic interference that may be dangerous with the change in climate by reducing excessive greenhouse gases (GHGs) production. Informed consent procedures prior to implementation give importing countries to prevent the importation of substances that have the potential to harm before they cause damage or harm to citizens or the environment (Atapattu, 2007).

2.4. THE POLLUTER PAYS PRINCIPLE (INTERNALIZATION OF COSTS)

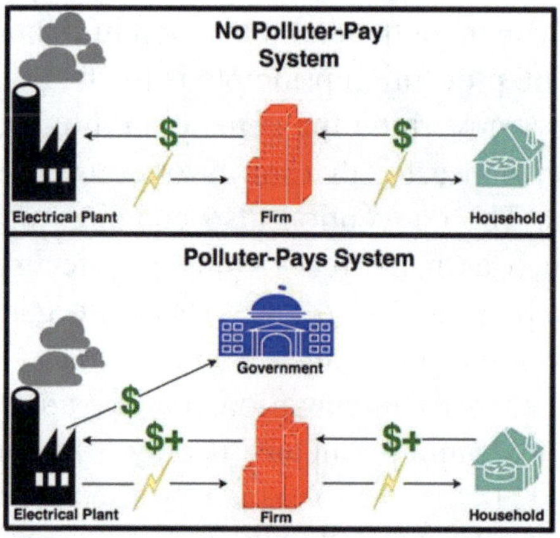

Figure 2.1: How the polluter pays system works and how it affects both consumers and producers.

Source: https://energyeducation.ca/encyclopedia/Polluter_pays_principle.

2.4.1. Description of the Polluter Principle

The polluter pays principle was created to make the parties responsible for releasing pollutants to pay for damages to the environment. It is regarded as a custom at regional level from the support it received by many organizations for economic cooperation and development and member countries of the European Union (EU). It is one of the most basic and important principle in environmental law globally (Atapattu, 2007) (Figure 2.1).

The principle integrates protection of the environment, development that is socially related and activities related to the economy by using instruments for regulation to ensure those responsible for pollution take the responsibility for environmental and social costs of their actions and that the cost referred to is reflected in the price of the goods and services at the market. The elements of the polluter pays principle are identified in Agenda 21 as follows the environmental costs in producer and consumer decisions to stop the tendency of treating the environment as a good that is free and to pass the costs to the society, countries, and other generations (Sands, 2003).

Elements of the polluter pays principle are also integrated in social and environmental costs. This is done so that prices are reflect on relative scarcity and the total value of resources and contribute towards preventing degradation of the environment.

2.4.2. Origin and History of the Polluter Pays Principle

According to the French environmental historian Jeane Batiste Feroz the polluter pays principle previously referred to financial compensation was already a principle for regulating pollution adopted by industries in the 19th century. The French historian wrote that the principle which is now offered as a new solution accompanied by the industrialization process was intended by the manufacturers themselves.

During the 1970s the OECD recommended that companies and industries should take full responsibility for costs that are related to environmental pollution control laws and governments can offset costs through subsidies. The cost of these environmental measures can be reflected in the cost of goods and services that lead to pollution during their consumption, production or disposal. OECD also encourages the use of scarce resources in the use of scarce resources and also to prevent instability in trade and investment globally.

2.4.3. Applications of the Principle in Environmental Law

The polluter pays principle is the basis for environmental policies such as an Eco tax. It is implemented by governments to stop prevent and reduce emissions such as GHGs. The principle is based on the fact that as much as pollutions can be sometimes unavoidable, the individual or industry responsible for the pollution must pay some money for the rehabilitation of the environment that is polluted.

European Union: The polluter pays principle also adopted treaty on the functioning of the EU and directive of the European parliament 2004/35/EC of the European parliament and of the council of 21st of April 2004 on environmental liability with relation to prevent and remediation of damage to the environment based on the principle. This directive entered into force in 2004. The member countries were given three years to implement the directive into their local laws and by 2010 the month of July all the member countries had done this.

France: Polluter pays principle is contained in the charter for the environment in France. The principle states that involved parties shall be required based on the provisions by law in order to contribute to the making good of any damage an individual, company or industries that may have brought harm to their surroundings.

Sweden: The polluter pays principle is also referred to as the producer responsibility extension. Thomas Lindquist first described the principle for the Swedish government in the year 1990. It seeks to shift the responsibility of pollution from the government to those who produce it. The cost of waste disposal is internalized into the product cost. In theory those producing the waste will improve the profile of the waste they produce, hence decreasing waste and increasing the reusing or recycling possibilities. The extended producer responsibility is defined by the organization for economic cooperation and development as a concept where importers and manufacturers of products are held accountable for impacts to the environment. This is throughout the life cycle of a product, including impacts inherent in material selection for products, the process of production and the impacts from the use and disposal of products. Producers or manufacturers have the responsibility to minimize environmental impacts while designing the product and also accept the responsibility of environmental impact that cannot be prevented by the product design. The producers are should accept legal, physical, and socio-economic responsibilities.

Switzerland: Polluter pays principle is the basis for managing waste in Switzerland. It is also applied in the collection of solid waste by municipalities for example waste bags are taxed by making payments pay per bag fees in three-quarters of the communes and since its adoption, the rate of recycling has doubled in a 20-year period.

United Kingdom: The operation of the polluter pays principle was established by regulations related to damages caused to the environment for England in 2009 and the environmental damage regulations in 2009 for prevention and remediation.

United States: The principle is employed in all major the United States pollution control laws for example the act for clean air, the act for conservation of resources and their recovery and the super fund. There are taxes underpinned by the polluter pays principles. Some of them include the gas guzzler tax for motor vehicles, a polluter pays fine for the corporate average fuel economy and the superfund law which requires polluters to pay waste clean-up that are hazardous from the sites of disposal if and when the said polluters are identified. A law amendment was passed in 2003 and was antagonistic to the polluter pays provision in the constitution of Florida.

2.4.4. Limitations of the Polluter Pays Principle

An observation has been made by the Environmental protection Agency (EPA) that the principle is not fully implemented in laws and programs by the U.S related to the environment. There are limited mechanisms to assess those who cause pollution in order for them to pay for the cost of treatment at the same time water for the use of drinking and treatment of sewage are subsidized.

India: The Indian Supreme Court interprets the polluter pays principle as absolute liability for harm caused to the environment. The costs not only include cost implications for environmental degradation but also extending costs to compensating the victims of pollution. Sustain able development also includes remediation of the environment that is damaged.

Ghana: Polluter pays principle was adopted in Ghana in 2011.

Zimbabwe: The 2002 Zimbabwean act for managing the environment prohibited the release of pollutants to the environment. The act requires people to the meet the cost of a polluted environment being decontaminated.

2.4.5. The Polluter Pays Principle in International Environmental Law

The polluter pays principle was mentioned in relation to environmental law at international level in the principle 16 of the Rio declaration which occurred in 1992 on environment and development. The declaration from the Rio conference states that authorities at the national level should work to promote environmental internalization of costs and economic instruments use, it also takes into account the approach individuals or companies polluting the environment should use, according to the principle the polluter should take the responsibility of pollution costs, with regards to interest of the public and without causing instability to trade (Banks, 2018).

The principle of the polluter pays is integrated in the preamble section of the Stockholm convention. It recognizes need for application and implementation of the polluter pays for the polluter principle as described the principle 16 in the Rio convention. In May 2000 in the first global ministerial conference environmental forum the adoption of the polluter pays principle was reasserted in the Malmo ministerial declaration.

2.4.6. Problems with the Polluter Pays Principle

There are different problems that are encountered in the implementation of the polluter pays principle, for example, the methods of correction are more effective than others as much as the government will encourage the reduction of waste or paying for damages caused, most of the time it does not provide means make the pollutants less harmful. It is also not easy to measure the amount of waste produced by a single company in the event it is released to the environment and there may also be a lack of transparency on the part of the company. Permissible amounts of substances released to the environment differ among countries and the taxes may differ which leads to one country benefiting over another. It is also difficult for countries sometimes to comply and agree to similar levels of pollution due to differences in economic development.

2.5. THE INTEGRATION PRINCIPLE

Environmental protection requires that due consideration should be given to potential consequences and decisions that will affect the environment. Various organizations and countries have made consideration associated with the environment into their process of decision making they include the EU, the United States, and business organizations such as the chamber of commerce. This has been done through mandates such EIA and other provisions (Sadeleer, 2002).

Application of the integration principle was found in the 13th principle of the Stockholm declaration in 1972 and 4th principle of the Rio declaration on environment and development and was further developed at the 2000 world summit on sustainable development the declaration on sustainable development in Johannesburg. In 2009, the EU had a directive on renewable energy with one of its goals as sustainable development of biofuels which is a controversial source of energy. It requires the countries involved to reach as certain quota of renewable energy and half of it is required to come from the sector of transportation. It also establishes the requirements to ensure the fuel sources are produced in a way that is sustainable.

The principle of integration requires integrating the protection of the environment, development of the environment and human rights at the level of developing concepts on policies and laws at the time of their implementation. True to the statements made internationally on sustainable development, the requirements for adopting the directive reflect the integration principle by trying to integrate development of the economy related to the industry of biofuels with concerns related to the environment and to some extent the rights of humans. The mechanisms take the criteria or form of sustainability required to be followed by members, monitoring, and reporting duties given to the European Commission. The principle through the directive manifests itself theoretically at both the stage of conception and implementation. The biggest concerns come from the globalized nature of biofuel and feedstock trade and its application in states that are not part of the EU. There are several questions, whether the directive will be sustainable without the governance of EU members at the international level (Sadeleer, 2002).

The integration principle can be constructed as an objective. It requires laws being developed in order to fulfill it. It can be therefore being seen as a means of achieving the policy on integrating the environment and beyond that achieving sustainable development. The integration principle can be interpreted as a rule of reference and it can also be seen as a principle that is autonomous (Sadeleer, 2002).

2.6 THE PUBLIC PARTICIPATION PRINCIPLE

2.6.1. Description of the Principle

The principle of public participation identifies as being essential to governments that are accountable, it also brings up environmental concerns and the concerns of organization in general. The United Nations environmental program presents public participation and transparency as needing protection in a manner that is effective. The rights of humans to hold and express their opinion and also seek and contribute ideas should be protected. Humans also have a right to access information that is right, comprehensible, and in a manner that is timely. This information is held by the government, organizations or industries on both economic and social policies which are related to use of natural resources to ensure their sustainability. The principle is also applied or presented in EIA, laws that also require publication and access to environmental data that is relevant and also administrative procedures (Koivurova, 2013).

It is therefore the duty of all people to participate in decision making in relation to activities related to the environment. Public participation areas vary from use of forest resources, soils, minerals to wildlife for consumption and non-consumption purposes. Other issues and concerns are related to solid waste for example garbage collection and disposal and demolition materials and hazardous wastes. The third issue is activities that lead to pollution of the environment.

2.6.2. Application of the Principle of Public Participation

According to article 19 of the universal declaration of human rights of 1948, every individual has a right to freely express their opinions. The rights include the freedom to hold opinions without there being interference and also seek, receive, and impart ideas and information through various channels such as the media and other frontiers (Koivurova, 2013).

At the United Nations conference on human environment in 1972 declarations were made in relation to public participation. In order to achieve global environmental goals, there is a demand for the acceptance of responsibility by communities and citizens and by enterprises and institutions at every level. All the parties are required to equally share the effort. The environment of the future can be shaped by different individuals and organizations from different fields.

Another declaration was that man has the basic rights to freedom, equality, and proper life conditions, in a surrounding that allows a life of dignity and well-being on the other hand humans should also take up the responsibility to protect their environment presently and for future generations (Koivurova, 2013).

The principle of public participation is also included in the Rio declaration on environment and development in the year 1992. Agenda 21 and principle 10 of the Rio declaration established a framework for participation of the public in various agreements that are multilateral which were adopted at and after the conference at Rio. The nations participating agreed to adopt the principle in relation to issues that deal with the environment. Issues related to the environment are best dealt with by allowing the participation of all citizens that are concerned at all relevant levels. All individuals shall have access to information at the national level which are associated with the environment. The information is held by public authorities some of the information related to hazardous waste and activities in the community and also having or been given the opportunity to take part in the processes of decision making. States are also required to facilitate the awareness of the public and participation by making information available to its citizens. Its citizens should also have access to proceedings that are judicial or administrative including redress and remedy shall be provided and this is the tenth principle of the Rio declaration (Koivurova, 2013).

It has three pillars that is information access, participation by the public and finally the access to justice.

- Access to information includes there being openness and transparency in making decisions and also making environmental regulations more efficient and effective. Authorities should make sure that the information is provided in manner that is transparent. The most important component of the pillar is the right to know mechanism.
- Public access to making decisions -includes public participation that is well informed, integration of citizen concerns into making public policies. A government's ability to respond to concerns by the public should be enhanced. The public should be notified about processes of decision making in the early stages and also in a way that is adequate. Transparent provisions should be made for parties that are interested in the process of decision making.

Groups or individuals should be allowed to submit information, comments, and analysis that is relevant environmental issue at hand. Other areas include demands made by the public, improving the compliance and acceptance of environmental decisions that have been made. These actions will build and strengthen the ownership of responsibilities to the environment by enhancing conformance to environmental laws and regulations.

- Finally, the third pillar is the access to justice which means groups or individuals have opportunities to address issues or complain when relevant authorities fail to comply with the duty of providing access to relevant decision making and information access. This pillar itself is expressed more in national laws rather than those adopted internationally. It also incorporates the protection of human rights which enable them to gain access to information and also participate in the decision-making process. The citizens in a country should also be allowed to question and challenge decisions made by their governments. Legal Bodies that possess a level of competence should be allowed to protect environmental rights. There should also exist processes that are independent and expeditious to deal with the damage to the environment.

The Aarhus convention on information access, public participation in making decisions and access to justice related to matters of the environment providing the most comprehensive articulation of the tenth principle which was found in the multilateral binding agreement. It also provides the member states with tools and standards for performance monitoring and engagement in decisions made in relation to the environment. It also applies the rules by which countries allow participation of the public in the process of governance at national level. It also required the states to apply the principle in any agreements to which they take part in (Beder, 2013).

All major multilateral agreements that include chemicals provide opportunities that are significant for the public to have access to information on the decision making process and also their proceedings. For example, the Stockholm convention required involved and participating parties to promote and facilitate education and public awareness especially for women and children on organic pollutants that are termed as persistent as well as the health and environmental effects of the pollutants and giving alternatives if there exist any. These proceedings and meetings allow the participation of nonparty observers (Beder, 2013).

Allowing the public to participate in decision making can be quite costly and time consuming but many benefits can be reaped from it. To begin it is a fair way of governance as it is based on the fact that those who are to be affected by decisions should have a say in it and should therefore they should have a right to determine the direction of the outcome. Secondly, participation gives dignity to those affected and involved. It helps to build the development of awareness and capabilities of involved parties and those affected by decisions. Thirdly it leads to making decisions better as it acts as an aid to those who make policies as participation will bring out public needs, feedbacks on issues, problems, and concerns of the public (Beder, 2013).

2.7. ACCOUNTABILITY AND TRANSPARENCY PRINCIPLE

The rule of law refers to the principle of governance which reflects the growing awareness of the importance of a government that is honest, accountable, and transparent in order to achieve sustainability of development as well as an increase in the understanding of the effect of corruption of morale to the public, also increasing economic efficiency, stability in politics and sustainable development in general. It is central to sustainable development. It integrates environmental needs with essential elements of the rule of law, it provides the basis for governance in relation to the environment. Good governance implies that governments and international organizations should adopt decision-making procedures that are transparent at the same time democratic and also financial accountability. They should also respect the due process in their processes and procedures and also observe the rule of law. Human rights should also be protected and finally, functions that are public should be conducted in a manner that is transparent and non-corrupt such as the implementation of laws and regulations, public procurement and customs (Bosselmann, 2016).

Good governance implies that citizens or individuals have the right to be fairly treated by states and intergovernmental organizations and also actors that are non-state and this may include businesses and non-governmental organizations. They should be subject to appropriate democratic governance internally, financially reporting that is accurate and accountability that is effective. In addition to this good governance calls for responsibility socially and investments that are socially acceptable and responsible as conditions for the existence of a global market that is sustainable.

2.7.1. Fundamental Principles of Good Governance

Sustainable development can be achieved through good governance which can be achieved in a way that is fair and effective. The principle of good governance includes:

- Participation: this can happen through the use of representatives or intermediate institutions that are legitimate. They a responsibility to make information accessible.
- The rule of law: it requires a framework that is legal in order to enforce judicial powers in a manner that is impartial and cannot be corrupt.
- Transparency: it means the enforcement of rules and decision making process follow rules and regulations. Information should also be accessible to those who will be affected by the decision making process and the enforcement. It should also be provided in a form that can be easily understood through different channels.
- Responsiveness: institutions should be able to promptly respond to the environmental concerns of its citizens and concerned parties.
- Accountability: it a key requirement in good governance. The relevant institutions should be accountable to those affected by their decisions.
- Equity and inclusiveness: it not only serves the interests of the mainstream society but also groups that are vulnerable.
- Effectiveness and efficiency: governments and relevant institutions should try to meet needs of the society and at the same time ensure they do this in a manner that is sustainable.
- Consensus oriented: different interests within the society should be taken into account in the process of governance and similarly the decisions following should try to reach a wide consensus on what puts the interest of the community at heart.

2.7.2. Origin and Application of the Principle of Good Governance

The principle is becoming increasingly important for use as an international instrument. The United Nations millennium declaration and the sustainable development goals contain a language of good governance. The fourth paragraph of the WSSD plan of implementation states that: the practice of good governance at country and international level are of importance if we are to achieve development that is sustainable. The basis for sustainable development at state level include good environmental, social, and economic policies; whether institutions respond to people's needs in a way that is democratic; the rule of law; whether there are enabling measures against corruption.

Recent years have also experienced a number of treaties that deal with corruption including treaties which have been negotiated under the United Nations auspices, the OECD, the European council, the American states organization and regional and sub-regional organizations (Bosselmann, 2016).

The explicit mention of good governance is not contained in many global chemicals but it is still contained in provisions about trafficking that is illegal which can be a problem especially for countries that have a weak form of governance. For example, the Basel convention contains provisions that are detailed with regards to transboundary trafficking of wastes that are considered hazardous including the requirements of those involved and this includes legal, administrative, and other measures that are appropriate to take action against trafficking that is illegal (Atapattu, 2007).

The Rotterdam convention notes the provisions of agenda 21 that are pertinent with regards to illegal trafficking prevention of toxic and dangerous waste at international level and it also it recognizes the importance of the standards in the international code of conduct and the united nations environmental program code of ethics on the trade-in chemicals at international level. Moreover, in the call for transparency and participatory chemical management processes, the chemicals rely on the principle of good governance among those interested.

The rural of law is a concept that is important to the United Nations organization and it defines the rule of law. They are many environmental pressures that exist from climatic change. Loss of biodiversity, scarcity of water, pollution of water and air, degradation of soil, and other contribute to the social inequalities that continue to grow. Environmental problems are intensified by conflicts over natural resources. Those vulnerable to environmental degradation include poor people, women, and children. Natural resources can be properly managed in a sustainable manner through being transparent and using the rule of law as a basis and therefore can be an engine for sustainable development as well as a platform for peace and justice. The rule of law is also essential for equity (Atapattu, 2007).

In 2013, in the United Nation Environmental Program's first session universally in 2013. The governing body adopted decision 27/9 on advancing justice, governance, and the environmental sustainability law. It was the first document negotiated internationally to establish the term environmental rule of law. It invited governments and relevant organizations to combat noncompliance with environmental laws by reinforcing cooperation at different levels in terms of regions.

The member countries recognized the importance of this principle in order to reduce violations of laws with regards to the environment and overall achieve development that is sustainable and its violation may undermine the process of achieving sustainable development. They also agreed that the implementation of goals related to the environmental at both national and international level that the rule of good governance was essential in the reduction of such violations. Since 2013, the legal division has housed the UNEP's environmental rule of the initiative of law (Sands, 2003).

2.8. EQUITY AND EQUALITY PRINCIPLE

Figure 2.2: A diagrammatic representation of equity.

Source: https://thefactfactor.com/category/facts/law/civil_law/environmental_laws/.

2.8.1. Description of the Principle

It is defined the UNEP to include intergenerational equity which is the right for generations in the future to enjoy a level that is fair from generations before them and intragenerational equity is the right of people within their generation to have access to resources that are there in their present generation. Environmental equity makes consideration of the generation that exists currently but under a responsibility to be accountable for long-term impacts of their activities and also act to sustain the environment globally as a base of resources for generations in the future. The control of pollution and management of resources related laws is assessed against this principle (Figure 2.2).

Growth of the economy can only be sustained when it is based on the equity and equality principle. Equity means that there should level of income at which people should meet at a minimum and quality of the environment below which nobody falls. At the same this means people should be able to gain access to resources but no individuals or groups should be required to hold the burden of the environment more than others as a result of actions by the government.

2.8.2. Concepts of the Principle of Equity

The terms equity, fairness, and justice are sometimes used in place of each other although they have differences that are subtle. We can understand the concept of equity from the following factors:

- There should be equal opportunity for one to achieve their potential.
- Benefits should also be equally shared by relevant stakeholders.
- Benefits of a particular public policy and cost should be fairly distributed, public funds should also be fairly allocated and this includes natural resources.
- Positive discrimination and redistribution to rights that are wrong historically or should in favor of groups that are disadvantaged economically, socially, in relation to gender and societal positions.
- There should exist equity among nations or internationally which should operate in the world of relations that are societal.
- Equity should go beyond national boundaries that is in terms of gender, indigenous community membership or existent vulnerable community.

The above aspects should be addressed in order to achieve development that is sustainable. If we emphasize on only one or some of the aspects the principle of equity becomes distorted. The concept also demands that attention is additionally paid to inequities in history and also discrimination. Others include how resources were initially distributed; rights and opportunities. When equity and sustainability are brought together social relations are transformed, rights, and resources are distributed, and policy approaches are also transformed in a holistic manner and simultaneously.

2.8.3. Applications of the Principle

The objectives of the principle help in meeting needs that are current without compromising the future of generations to come. It also gives provisions for the fair treatment and respecting the rights of resources in the environment that are non-living. To persuade individuals pursuing their livelihood not to cause harm to other people and also the environment. Another objective is to guarantee access that is equitable to natural resources that exist in the environment. Safeguarding those that are vulnerable in the process of development and finally guarantee allocation of resource rights in a manner that is fair so that resource management can be done collectively and in a cooperative manner (Beder, 2013).

The Brundtland commission played an important role of making the concept of equity popular and together with sustainable development and it was subsequently reiterated in their report in 1987 our common future which was similarly supported by the United Nations and the definition was globally adopted. The concept of intergenerational equity was adopted at Rio conference in 1992 as a framework for the improvement of the quality of life at global level and has since then been include in many strategies for sustainable development. It defined equity as also being fair. The concept of equity is derived from social justice. The belief it represents is that there are things people should gain access to which are mostly their basic needs and that burdens and benefits should not be widely spread within and a cross individual within a community; policies made should be implemented and directed in a manner that is impartial fair and just at both ends.

A central theme in development that is sustainable is the integration of the economy together with social and environmental concerns. This reasoning is found in agenda 21. In the action plan for development that is sustainable and was signed by more than a hundred at the summit. It possesses three basic objectives to incorporate environmental costs in decisions of consumers and producers, to move towards the integration of social and environmental costs into income-generating activities and finally to include market principles in the process of framing of economic instruments whenever it is appropriate. Agenda 21 also brings forward the need for integration of accounting related to the environment and economy in procedures for national accounting (Beder, 2013).

Many approaches are taken up by government or authorities to achieve sustainable development sometimes worsen the inequities that already exist in some of the countries. Equity is supposed to be the central of sustainable development. These inequities that exist can merely be solved in theory but is not really the concern of the environment or either the said governments do not have the understanding of the consequences of inequity brought by policies that are promoted by people or groups with different agendas and priorities.

Inequity is a problem that should be taken seriously and new ways of making decisions must also be established to ensure the consideration of different values related the environment.

2.9. SUSTAINABLE DEVELOPMENT PRINCIPLE

2.9.1. Description of the Principle

Sustainable development is defined by the UN Environmental Program as development that helps people to meet their needs presently without affecting the capacity of people to meet their needs in the future. Concepts that are also mentioned in relation to development that is sustainable include integration, interdependence between social and economic development and protection of the environment. This principle can be used to assess laws that mandate EIA and those that require development or encourage it in a way that minimizes impacts on the environment (Bosselmann, 2016).

The sustainable development concept that is modern was a discussion topic at the 1972 at the Stockholm conference and was still incorporated in 1983 at the world commission on environment and development or the Bruntland Commission. At the earth summit in 1992, the Rio declaration resulted in the Principle 3 which states that:

"The right to development must be fulfilled so as to equitably meet developmental and environmental needs of present and future generations."

The sustainable development principle to harmonize the conflict between industrial, economic, and social development in order ensure the environment is health. Many environmental studies that have been done have shown that most environmental problems are as a result of technological and industrial development in developed countries and in developing countries it is a result of poverty, overpopulation and illiteracy. Progress in the development of an economy is at the best interest of countries but should not be at the cost of the health of the environment because it not only affects the current generation but also generations to come. Sustainable development is therefore needed in order to maintain a balance between environmental protection and development. The principle of sustainable development emphasizes on two basic needs, first the need to achieve development that is socio-economic and the second is the need for limitations to be imposed on the coping ability of the environment in order to sustain them for future generations (Bosselmann, 2016).

Discussions on environmental issues have the concept of sustainable development as their key concept ever since including the earth summit of 2002 and the United Nations conference of sustainable development which was in 2012.

2.9.2. Objectives of Sustainable Development

The basic objectives of sustainable development include:

- To maintain goods and services production for both development and efficiency.
- To manage and conserve neutral resources including bio-diversity preservation and maintaining biological integrity.
- Enhancement and maintenance of the quality of life and adopting the principle of distributing wealth and material resources in a manner that is equitable.

They are referred to as economic, environmental, and social objectives of sustainable development. The objective of the principle of sustainable development from an environmental point of view is centered on three main issues which include the maintaining process of ecology that exist in the environment, preserving the diversity of genetics and securing sustainable species use and the use ecosystems.

2.9.3. Application of the Principle and Progress

Sustainable development involves various approaches economic, human, environmental, and technological. The process seeks to improve human life quality along with the conservation of the ecological system. This makes development and the environment interdependent. There cannot exist development without environmental protection and vice versa (Bosselmann, 2016).

When agenda 21 was drafted principles were introduced in order to meet the objective of sustainable development and they include intergenerational equity, natural resources conservation, protection of the environment, the precautionary principle, the principle of polluter pays, the principle of liability, eradication of poverty and the principle of public trust.

Lack of progress in turning agenda 21 into account led to another summit in Johannesburg in 2002 on sustainable development and a thrust was put on forming partnerships for sustainable development some of which failed to be implemented. Before the summit leaders had taken the initiative to develop millennium development goals which were set to be achieved in 2015 of which environmental sustainability was one of the problems identified as from previous summits. In 2007, it was indicated that the MGDs would not be realized by the year 2017 as resources failed to be mobilized in order to achieve the goals of millennium development.

In 2012, the United Nations met to discuss sustainable developmental goals after the failure of achieving the millennium development goals. A list of 17 items were developed. Sustainable development acknowledges that growth should include a sound environment in order to achieve a reduction in poverty to sustain natural resources for generations yet to come. The efficient use of resources would benefit the current population and also benefit generations in the future. This would also include proper planning to benefit both the people and the environment. The three pillars of sustainable development still remain economic growth, stewardship of the environment and also social inclusion which is carried across all development sectors (Bosselmann, 2016) (Figure 2.3).

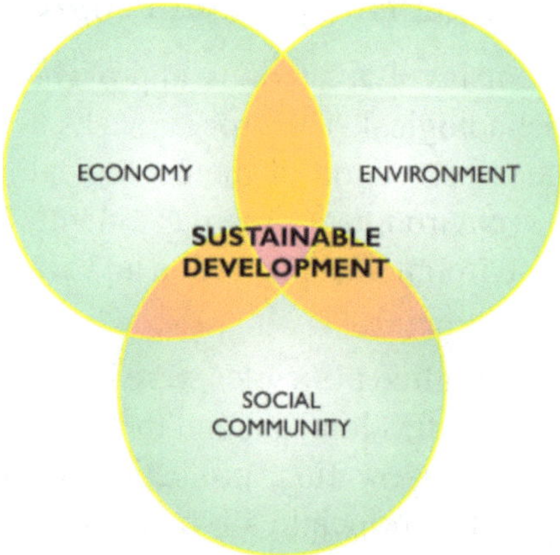

Figure 2.3: The three pillars of sustainable development.

Source: https://www.sciencedirect.com/topics/earth-and-planetary-sciences/sustainable-development.

The concept of environmental protection is incepted to foster the achievement of development that is sustainable and it is the third pillar and is a concern of many. They are different measures that have been put in place by different countries in order to sustain the environment including the concept of the 4Rs which is reduce, recycle, recover, and reuse. Other include measures of environmental conservation wildlife, water resources, forest resources, sustainable land management and the management of waste. Protection of ecosystems, the quality of air, integrity, sustainability of resources and elements that bring stress to the environments are aspects that are also looked into when protecting the environment. Another concern is how technology will be used to drive generations into a greener future; technology should also be used as a key to achieving sustainability why still ensuring that the environment is protected for future generations. Indicators can be used to see how far a nation has come in achieving sustainable development (Bosselmann, 2016).

2.10. TRANSBOUNDARY RESPONSIBILITY

It is defined in international law as the responsibility to protect the environment of individuals and at the same time prevent damage to the environment to that of neighboring countries. The United Nations environmental program considers transboundary responsibility at international level as limitations to a state's rights. Laws that are made to limit externalities imposed on the health of humans and the environment are sometimes assessed against the transboundary principle (Sadeleer, 2002).

The principles of international law, imposes liability on the acts that are illegal by individuals or groups. The principle is reflected in the articles on responsibility of the state by international law in 2001. The terms of liability for damage to the environment are still evolving. Liability for damage has been included in treaties and other international acts. It has four elements fauna, flora, soil, and water; material assets; amenities of landscape and the environment and the interrelationship between the different factors. Liability serves both national and international purposes and may be used as an economic instrument to foster compliance to environmental laws and regulations. The principle can also be a basis for imposing sanctions in the event of wrongful conduct or in case corrective measures are needed for environmental restoration. It also provides a technique for the internalization of both environmental and social costs into the process of production. Principle 22 from the Stockholm declaration recognized gaps in relation to liability and called for the need to develop laws and regulations with regard to the same that is the liability and compensation of those affected by environmental degradation such as pollution for victims affected within and beyond the jurisdiction of a country (Sadeleer, 2002).

3

Importance of Environmental Law

CONTENTS

3.1. INTRODUCTION AND HISTORICAL DEVELOPMENT OF ENVIRONMENTAL LAW

Environmental law is also known as environmental and natural laws is a collective term that encompasses legal aspects towards the protection of the environment it is a distinct and very specific set of regulatory principles that goes towards the correct functioning management and governance of natural resources like trees, fores,t oceans, and minerals. They pertain to anything from principles to policy directives and regulations and functions as guidelines on how human beings relate to the environment covering a very vast range of topics many of which we will explore in this chapter.

The term environmental law does not just entail government law but also covers businesses and organizations that try and set up a regulatory framework that is not legislative per se. It translates to acting responsibility and ethically with a good understanding towards the environment and ecological systems we live in with an overall comprehensive realization of our impacts. From a practical and realistic perspective whichever way you look at it environmental and natural resource law has an effect on all of us individually with our health and collectively with our business activity and overall sustainability. In this chapter, we will not only take a look at the environmental laws that Govern us but also why environmental law is important (Banks, 2018).

In as much that human beings have been in contact and aware of our environment far longer than they have been laws to protect the said environment, the late twentieth century has provided and developed environmental law farther than just a modest adjunct of regulations towards an almost independent field in itself. This has been extremely beneficial not only in protecting the non-human nature but in also preserving human health. Historically national governments have intermediately passed laws to protect human health and curb nuisances which in common law and tort law is causing a substantial and unreasonable interference. In Rome for example the Senate of Rome worked on regulations and legislations to protect the supply of drinking and bathing water that enters the city. Such laws allowed for action against damages or injunctions if there was damage done to land and so measures could be carried out against owners of pig sites if there are Unbearable smells emanating from it or strict liabilities could be set against dumping of rubbish. The earliest legal enactments to consciously preserve the environment were difficult to be imposed because the private enforcement was limited and inadequate the dealings of large environmental threats especially to common resources. Over time as pressures mounted legislations became more stringent and wider practices were administered for example in the 14th century England passed a law prohibiting the Burning of coal in London and the disposal of waste into waterways. This is a perfect example of the failures in private enforcement because by 1858 central London suffered from the Great Stink which was an incredibly terrible smell from untreated human waste and industrial effluents into the river Thames that was exacerbated by the hot summer weather. The situation was so bad that three major outbreaks of cholera were reported that can be directly attributed to the Great Stink. Parliament had to be evacuated in the summer heat because the smell was so crippling and the fear of miasma which is a theory that States the disease are spread from Bad noxious air or bad air quality (McAllister, 2008).

More actively preserves legislations include the 1681 order by William Penn the leader of the English colony of Pennsylvania stating that for every five acres cleared for settlement one acre of Forest is to be preserved. Benjamin Franklin also carried out several campaigns to minimize the dumping of waste in an unregulated fashion. The 19th century brought about the Industrial Revolution with an unregulated immense burning of cool and chemical manufacturing that negatively impacted public health and the environment. These early analogs of environmental law were conceptual at best with environmental law becoming a distinct body of its own later in the 20th century. Before that even international environmental agreements were non-existent and the listed accords focused primarily on shared waterways and their boundaries ignoring pollution control and other ecological issues until, just like the great Stink, it was too little too late. 20th-century developments realized and recognized the natural environment as a fragile entity needing special legal security conservation and Preservation.

This is where the rise of conventions to protect commercially viable species like the convention for the Protection of Birds useful to agriculture will signed by European governments for example 4D convention for the Protection of migratory Birds signed in 1916 adopted by the United States. These concepts spread from Europe and the United States to countries in Africa for example Egypt South Africa and Sudan adopting the convention relative to the preservation of fauna and Flora in their natural state that birthed many national parks and reserves. They were mainly ratified by the colonizers on behalf of the State that they governed.

Environmentalism which is the philosophy and ideology of Environmental conservation breeding into a political and intellectual movement was popularized in the 1960s especially in the United States with the publication of Silent Spring by Rachel Carson that examined chlorinated hydrocarbon pesticides and the detrimental effect on the environment opening up discussions on potential environmental hazards across-the-board. These discussions included addressing issues concerning waste disposal air pollution and endangered species that fostered the creation of the Environmental Protection Agency (EPA) to continue maintaining compliance levels on a federal level (Banks, 2018).

In Asia more specifically Japan for example the Second World War ending was paired culturally and economically with the Rapid industrialization and laws were needed to govern the unregulated release of industrial and chemical waste, especially into the food chain in some areas. Minamata a city in Japan so many people suffering from mercury poisoning after eating fish that were contaminated and the Japanese government had to develop pollution control policies that were comprehensive against environmental pollution.

The second world war did not only have environmental effects in the industrialization of Asia but in the development of the United Nations who initially held the conference for the human environment in Stockholm in 1972 that established the United Nations environmental program that has been instrumental as a principle international environmental organization. They are responsible for the oversight of important conventions like that on the international trade in endangered species of wildlife Fauna and Flora but have little power in terms of imposing or enforcing sanctions on countries and governments that do not comply with agreements. The European Union (EU) and community have been slow in imposing legal standards for Environmental Protection however the environment Acquis is the somatic body of legislation that runs and imposes European law that focuses on environment. In Germany for example they were instrumental in changing public attitudes especially during the bad acid rain period in the 80s where the environmental German Green Party was founded and won representation in the Parliament prompting more stringent Environmental regulations for protection (Rostron, 2013).

From a historical point of view, the Kyoto protocol is also extremely important to mention and it adopted the framework for the convention on climate change especially in establishing emission targets and setting mandatory limits on greenhouse-gas emissions. It was immediately ratified by 55 countries and was to be put into effect in 2005 with the first commitment period expiring by 2012

3.2. TYPES OF ENVIRONMENTAL LAWS AND THEIR IMPORTANCE

3.2.1. Command and Control Laws

Majority of Environmental laws fall into a group of laws known as command and control. These laws have three descriptive facets listed below:

- Identification of environmentally harmful processes and activities;
- Imposing of conditions, standards, and regulations on these activities and processes; and
- Prohibition of any processes or activities that for short of compliance levels with imposed conditions.

Ideally all pollution control regulations fall under command and control laws that determine air quality water quality waste management chemical safety and contaminants disposal. In air pollution and equality laws emissions of harmful pollutants that are gases in nature into the atmosphere is governed. There is a subset of equality laws that dictates the quality of the air inside of buildings ensuring that equality is not compromised indoors as well as outdoors and a specialized to protect human health by lowering or altogether eliminating air pollutants (McAllister, 2008).

Air Pollution under Command and Control Laws in China and the United States: Great example to offer would be China and the development over the years where initially pollution was a so-called non-issue and the focus was on economic systems and building their strength. We lack of pollution regulation both ideologically and on practical levels was also exacerbated by China's political isolation and the Chinese people's immediate concern on the basic necessities of life such as food and clothing. The United Nations conference in Sweden that led to the development of the United Nations environmental program in 1972 lead to China's first form of Air Pollution control project where emission control was established through a pilot program in Shenyang. Bureaucratic structures started to develop henceforth with a group of 20 under the state Council or put in charge of Environmental Protection and provincial units of a similar nature were formed. No significant changes were made until the mid-90s and annual concentration levels of total suspended particles which is essentially an archaic regulatory measure for particulate matter in the air and sulfur dioxide levels remain the same violating world Health Organization recommendations several times over the limit (McAllister, 2008) (Figure 3.1).

Figure 3.1: Masks are now an ordinary and necessary feature in Chinese dress code when stepping outside in major cities.

Source: https://www.weforum.org/agenda/2019/06/chinese-professor-explains-what-air-pollution-does-to-your-health/.

The People's Republic of China air pollution prevention and control law was revised in 1995 after initially being enacted in 87 and is their main specialized environmental law along those lines the target industrial facilities and their emission standards. This is a type of command and control law as the national and local Environmental Protection bureaus have waste discharge standards that are meant to be complied with and they are formerly stipulated. The law also enacts a polluter pays principle and fees are charged to companies or industries together with pollution levies on those who exceed emission standards. In law a levy is an item ordinarily money that is seized to satisfy a legal judgment. In this way Enterprises have to give money to the government as a way to satisfy the environmental legal requirements. New facilities are required to exhibit compliance levels in their design as well as it being a condition in which construction permits are issued.

In this sense when discussing environmental laws their importance and the roles that they play we can see that the Chinese downplayed human health in relation two environmental laws for the sake of economic prosperity. It is a bad trait of considering that the economic power comes from the human workforce and their health was put at risk. The improvement has been made however especially with the pollution lover system where companies or organizations and Enterprises are charged a fee based on their release of airborne and even waterborne pollutions. There are still enterprises that do surpass and violate the standards on emissions and effluence and higher fees are placed on them (Banks, 2018).

The second stage in command and control laws concerning pollution as stated above dictates the importance of imposing conditions and in this way monitoring compliance is a vital element in implementing the pollution levy system. The Environmental Protection Bureau is authorized through article 23 of the People's Republic of China air pollution prevention and control law to design standards of monitoring and set up monitoring networks on both local and national levels. The Environmental Protection Bureau is legally given out of 32 penalized Enterprises that do not meet compliance levels and to organize monitoring stations that consistently check up on polluters activities. The environmental laws here also serve that important job of providing jobs and an economic livelihood as well.

When you compare air pollution control in China and the US you will find that western industrialization occurred much earlier and therefore the setup of Environmental laws and bodies to monitor and impose regulations are much more concrete. One way you will find this is through the monitoring of waste releases that is done through self-monitoring and self-reporting where the enterprise reports results to the environmental protection Bureau through their own monitoring systems where data is often falsified. In the United States, self-monitoring and self-reporting is a practice however permit holders that falsify any results are severely penalized in both civil and criminal Courts. The main reason this can be attributed tourist bureaucratic processes having existed way longer in the US with the first pollution statutes in the US being formed in 1881 where is China developed theirs in the last three decades. By the 1980s there were 50 states air pollution control jurisdictions we're at 1 at municipal levels and 142 in County levels across the United States with the Federal Government having their own say in 1955 through the air pollution control act (Rostron, 2013).

The high level of China's air pollution that was initially downplayed will result in economic losses due to adverse health impacts. The government is there for confronted with a burdensome challenge of creating better policies and bodies to implement them. On a positive note it has been reported that boost the local environmental protection bureaus and the Chinese central government have expressed a strong will and are transparently motivated to clean the environment and they are making progress in changing policy designs that are not feasible or enforceable (Figure 3.2).

Figure 3.2: Heavy air pollution enshrouded Tiananmen square in Beijing.

Source: https://cen.acs.org/articles/95/i4/Peering-Chinas-thick-haze-air.html.

Water Pollution under Command and Control Laws in India and the UK: India was yet another country that participated in the United Nations conference on the human environment in 1972 and realized the importance of the preservation of natural resources. The Indian government formulated the water act in 1974 two years later for the prevention and control of water pollution through their constitution. India also faced the same challenges of creating the appropriate bodies and assigning responsibilities for the prevention and control of water pollution. They were able to officiate central and state pollution control boards that respectively investigated and assessed all issues concerning water quality and water pollution aiming to maintain and restore the "wholesomeness of our aquatic resources"

The act stipulates that pollutants are not to be discharged into large bodies of water not limited to but including oceans, rivers, and lakes and it is their duty of the state pollution control board to terminate such activity. The Enterprises that failed to meet compliance levels are liable for imprisonment of a range between not less than one year and six months to as long as 6 years along with monetary penalties (McAllister, 2008).

As much as the water pollution act and many other environmental regulations have been impressive in both number and structure they have not been a success as a reported 70% of the water in the country is contaminated with as many as three and four Indians contributing to this. The biggest reason is argued to be the luck of effective control power and legitimacy given to regulators on both the states and national level. The air act of 1981 in India was meant to monitor and prevent the increasing levels of pollution. The numbers here prove that evolution over time has fallen by about 17% between 1987 and 2007 but in contrast water pollution concentration over the same time show mixed results. The successes in air pollution management has been due to environmental regulations that made catalytic converters mandatory for vehicles and that reduced total particulate matter as well as sulfur dioxide concentrations in the air.

The water act and national river conservation plan together with a myriad of other regulations have been unable to effectively improve the levels of water pollution. Studies done showed that water pollution policies have a lower community-driven initiative and the air pollution counterparts and a big reason is because again no institutional body has been mandated with powers to implement water pollution policies successfully (McAllister, 2008).

The water crisis in India is so bad that brought break out whenever water tankers carrying safe drinking water driven into some areas with many claiming that is suffering from the worst water crisis in history. A reported 70% of India's water is contaminated contributed to 20% of the country's disease burden and only one-third of wastewater is treated meaning that raw sewage flows into every imaginable water source. In major cities across India like New Delhi and Bengaluru arsenic fluoride and other carcinogens have really contaminate the water some with nitrates and pesticides seeping into groundwater and by 2020 it will affect I've reported 100 million people. As one can imagine this is a great burden especially on the poor as the prices paid by them. From an economic perspective India would have to sacrifice 6% of its GDP if the water crisis is not resolved due to food insecurity revealing the importance of environmental laws on food and crop production (Banks, 2018) (Figure 3.3).

Figure 3.3: Sushila Devi, who lost her son and husband in a water dispute, poses outside her one-room shanty in Delhi, June 26, 2018.

Source: https://in.reuters.com/article/india-water-crisis/indias-worst-water-crisis-in-history-leaves-millions-thirsty-idINKBN1JV23W.

Comparatively in England and Wales the laws are much more stringent through the environmental permitting regulations 2010 viewed from the water resources act of 1991. Literally any and all water discharge activity should be stated under an environmental permit for exemption from the Environmental Agency. Environmental agency was established in 1995 and they sponsored by the department for environment food and rural affairs under the government of the United Kingdom. It goes without saying that the environmental agency is non-departmental public body that carries a lot of weight in environmental protection and regulation we decent Barking from both European directives that are illegal acts of the EU together with the United Kingdom's government. The structure of the environmental agency has two directorates under policy and process; one dealing with flood and coastal risk management and the other with environment and business. This highlights the importance of Environmental laws especially when mandated by appropriate bodies that carry weight being a key component in urban development (Rostron, 2013) (Figure 3.4).

Figure 3.4: Wastewater from a factory being directed into a river contributing to water pollution.

Source: https://cdn.britannica.com/33/157933-050-E5FFFD7C/waste-river.jpg.

Waste Management under Command and Control Environmental Laws in Turkey and in Germany: Turkey has hard waste management regulations that date back over a century. In relation to waste management there earlier slow came into force in 1930 under the municipality law and general health law number 1580 and 1593 respectively they included provisions and instructions only collection and landfilling of waste and gave directions on the measures to be taken so that human health is protected. On a local level the municipality is responsible for the construction operations separation and disposal facilities. Turkey also was one of the 53 signatories on the Basel convention which dictates the control of the trans-boundary movement of hazardous material although it does not address the movement of radioactive waste however it includes various provisions and Waste Management which Turkey adheres to. Under the regulation waste generators that produced in excess of 1000 kg of hazardous waste must have a temporary storage permit from the Ministry of environment and urbanization and even though generators that produce less of the amount stated are exempted from obtaining the permit they must still provide information about their worst waste quantity and type and comply with various legislative waste management systems as well making various declarations through the environmental information system under their professional directorate (Banks, 2018).

Together with the environmental permit that covers a Nation's environmental noise and general waste discharge, there is an environmental license that is given to the facility that applies for it that addresses the technical sufficiency of the facility. When critically looked at the country actually has a very strong legal framework. Although municipal waste management is legally under the responsibility of the municipal the National government has made efforts, for example, drawing up a master plan between 2007 and 2009 based on the recognition that there is a culture of waste disposal that is unsafe and uncontrolled and poses a high level of risk to the environment and two it's 70 million citizens. 1011 there is a reported one sanitary landfills play municipality accounting for roughly 3000 in total a very large increase from the 1990s that had only 90 controlled landfills ((Rostron, 2013).

A similar story as in the failing case studies presented above existing Turkey as it is not the legal arrangements that is in question however the deficiencies lie in the implementation of the laws. There is also a week technical infrastructure that exists to get away the institutional insufficiencies. This situation is only made worse by the fact that at a national level Turkey is still yet to come up with a federal level certificate plan on general environmental issues including climate change as well. The Turkish ministry of environment and urbanization claim that by 2025 they will be carrying out the EU landfill directive realizing the situation is dire. The Turkstat that is the Turkish government Agency that producers of official statistics on Turkey as the Turkish Statistical Institute reported that uncontrolled open landfill waste producers 58% of the total national methane emissions between 1990 and 2011 and up until 2013 that imposed no landfill tax.

What is important to note here is the impact of poor waste management. The biggest problem in the country stems from uncovered uncontrolled landfills where garbage is dumped and left to rot. This means that there is an increased level of greenhouse gases (GHGs) and as the diagram above shows uncovered landfills produce biogas and microbial diseases on top of being a potential source of flammable material. Turkish waste management system is financially flawed as it does not take into account the pollute pays principle which we will discuss later. Turkey is also ranked as one of the worst countries and recycling capabilities as it recycles just 1% of its total plastic while the other 99% ends up in landfills (Banks, 2018) (Figure 3.5).

Figure 3.5: Uncovered landfills remain to be potential sources of flammable biogases, carcinogen, and toxic waste, as well as microbial diseases.

Source: https://en.wikipedia.org/wiki/Waste_management_in_Turkey#Government_efforts.

Germany on the other hand is a country that has great success in the waste management processes especially in achieving a recycling-based economy that both conserves resources and reduces the adverse effects that municipal waste has on the environment. The central concern is to avoid any negative impact that we have by recycling as much of it as possible and environmentally friendly fashion. The legal framework is based on European laws in the EU regulations and directives where the EU regulations apply to each member states while the directives are meant to be modified international law buy each member states and in Germany, the first waste disposal act was adopted in 1972 where all disposal of waste activities are subject to the closed cycle management act (KrWG). This law was created to work on a multi-level waste hierarchy that is encompassing waste prevention, reuse, recycling recovery (like incineration for energy generation purposes), and waste disposal (Rostron, 2013).

A federal law under the waste management act is only regulated by regional States who have jurisdiction over certain aspects of waste disposal that are not already well overseen and regulated by the federal law in essence they at a state level only implement. At the municipal level, waste disposal law is on imposing collection and recovery legislation majorly by integrating household waste into the public system. The Germans look at any waste with the mentality that it must be possible to reuse this substance even if it is just a byproduct together with the consistent never-ending effort to minimize waste production in the first place (Figure 3.6).

Figure 3.6: Waste sorting facilities where waste is segregated are indispensable for waste management.

Source: https://www.umweltbundesamt.de/en/topics/waste-resources/waste-management.

When looking at waste disposal Turkey and Germany offer very interesting outlooks as Turkey has very well defined legislation that is not overseen at a national level for it to be implemented well enough at the local level. Germany on the other hand have the federal laws that are well-defined and only requires implementation at lower state and municipal levels with each member of society making a conscious effort towards sustainable living. The importance of environmental law is in this case presented by the element of sustainable living where in Turkey methane gases and general greenhouse gas emissions have increased within the last two decades because of the unregulated and uncontrolled landfill Warehouse in Germany who in the past suffered from acid rain that burnt up forests now on a better path to words environmentally sustainable lifestyles. This ensures that future generations have a healthy environment in which they will not only survive but thrive. Germany also shows why environmental law is important in its creation of an entirely new economy within the environmental sustainability platform like burning of waste to produce power and the multi-billion-dollar recycling industry. One can imagine how much money a recycling company would be able to make with the 99% of plastics that find their way into more than 3000 landfills and dumpsites that they have (McAllister, 2008).

3.2.2. Environmental Contaminant Cleanup Laws

Environmental cleanup laws are different from pollution control laws in the sense that they are designed to be in response to environmental contamination and govern the removal of pollution or contaminants from the environment and all its media like soil surface water etc. Toxic spill and leaks after they happen are quite damaging and laws need to be put in place to establish protocols for cleanup and emergency response as well as liability allocations. Other laws that may fall under this category include feasibility studies and investigations, remedial action and site assessment. In many countries response and cleanup regulations are authorized a standalone law while in others they act as parts of larger laws that are specialized under environmental mediums of pollutants ((Lazarus, 2008).

One environmental body or organization that has been successful in clean up enforcement of laws has been the EPA of the United States. The EPA's clean up enforcement program is one designed for human health protection first and finding those responsible for contamination of the environment to reimburse the EPA for its cleanup activities. EPA has several independent authorities that also aid in specific cleanup situations and one viable example given would be the super fund enforcement that was created after the Love canal incident.

The love canal disaster involved the Hooker Chemical Company that was a company that disposed large amounts of chemical waste in a canal that was initially built on the Niagara Falls for the generation of electricity but abandoned. The hooker Chemical Company continue to dump chemicals in the canal up until 1952 it became apparent Liberia was going to be developed into a small town and construction had begun. The site during his 10-year lifespan became a host for nearly 20,000 tons of chemicals that consisted of various carcinogens alkalines and chlorinated hydrocarbons from the production of perfumes rubber and synthetic resins. As much as the chemicals were buried 6 to 7.5 m under the ground in the early 50s, the area was literally a ticking time bomb. During the 1980s after the town had become heavily populated with schools and communities the municipal decided to build after buying the land from the hooker chemical company at a throwaway price experienced are particularly wet winter where the chemicals began to leak. A terrible odor cupboard the town and basements and children's playgrounds were flooded with chemicals. Eventually after several birth defects and many more cancer cases resulting from the leached chemicals the town was evacuated and the cleanup program was ordered (McAllister, 2008).

First as the case is mounted the EPA as the mandated authority carried out investigations and found that the blood tests of residents around the area showed high white blood cell count which indicates what is a precursor to leukemia. Other studies were carried out through the EPA liaising with the United States national research council where they found the major exposure was through groundwater that seeped into the basement and children's playgrounds. The super fund enforcement was standard authority through the comprehensive environmental response compensation and liability act of 1980 and the program is designed to clean up sites either on their own or negotiating with the companies or people responsible for the cleanup or remedial services. The love canal incident was the first entry on the superfund list and eventually, it was a success with the site having been declared clean and deleted off the list in 2004 (Banks, 2018).

The litigation and compensation direction that this narrative took was that the hooker chemical company which changed its name to Occidental were rude as negligent but not reckless and they agreed to be 129 million dollars in restitution after being rightfully sued by the EPA through the justice department. The superfund was paid back in full 101 million dollars that are spent on the clean up together with 28 million dollars in interest. Remediation actions included destroying of houses on the Eastern and western side of the canal and less than 10% of the families opted to remain with many of them carrying out their own lawsuits that were settled years after the disaster. The most toxic areas that included 16 Acres had to be reburied with plastic liner clay and more dirt and as one can imagine the work was quite extensive. Environmental contamination laws like the superfund amendments and reauthorization act of the toxic substance control act and others serve as great examples of control and command laws under the general Environmental legislation.

Environment contamination laws are important in this sense because those responsible for polluting the environment are held accountable. With them they would not be any regulations concerning response to disasters and they act as a safeguard in protecting the land air water and soil which are the mediums of the environment. Furthermore, various punishments including jail time in some very extreme cases are very good deterrents from negatively impacting the environment (Banks, 2018).

3.2.3. Environmental Chemical Safety Laws

Chemical safety regulations are intentionally designed for the management and Governance of chemicals in human activities especially man-made chemicals that are used across a wide range of industrial applications. They seek to regulate the potential pollutants themselves and handling of them thereof rather than media-oriented environmental laws. This includes the banning of chemical constituents and regulating pesticides and plastics. They're not all specifically for less than one category as some may regulate the transport of dangerous goods inflammable and explosive goods and large-scale management of dangerous chemicals while others made look at safety regulations in the handling of said chemicals (McAllister, 2008).

Great examples of chemical environmental laws on the international scale include the aforementioned basal Convention on the control of the trans-boundary movement of hazardous waste and disposal which was ratified by 172 parties that provides protection for human health and the environment against chemicals that have adverse effects either in their generation management and disposal. The Stockholm convention is another that was an international environmental Treaty aiming to restrict the production or effectively eliminate the making and use of persistent organic pollutants which are also sometimes known as forever chemicals. These chemicals are resistance to environmental degradation and by accumulate with potentially adverse effects on human health and the environment. They include chemicals like solvents and pesticides and some Pharmaceuticals that are man-made. Specific examples include chlordane enduring hexachlorobenzene toxaphene and polychlorinated biphenyls. All these products have characteristics that exert negative effects like the fact that they are able to travel far from the source but worse off is there bioaccumulation tendencies.

The Basel Convention and the Stockholm convention both hard the secretary-general of the United Nations as the depositary figure and perhaps even more famous legal efforts made to words environmental chemical legislature was the Montreal protocol. The Montreal Protocol was initially signed in 1987 entering enforcement in 1989 and has since undergone nine different divisions. It was a treaty on an international level designed to protect the ozone layer by the reduction in the overall elimination of production of such chemicals that were harmful to the environment (McAllister, 2008) (Figure 3.7).

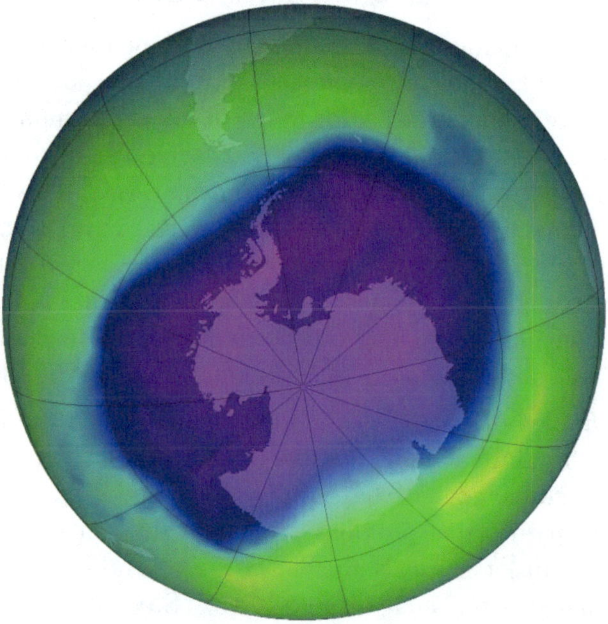

Figure 3.7: The largest Antarctic ozone hole recorded as of September 2006.

Source: https://en.wikipedia.org/wiki/Montreal_Protocol#/media/File:NASA_and_NOAA_Announce_Ozone_Hole_is_a_Double_Record_Breaker.png.

Chlorofluorocarbons and hhydrofluorocarbonsare the major group of chemicals that were found to deplete stratospheric ozone and generally all substances that contained either bromine or chlorine were deemed eligible to be phased out. Chemicals containing only fluorine do not harm the environment in the same way. The interesting thing about this law is that it was all in voluntary signatory level and countries like Russia continued to sell and produce significant amounts of chlorofluorocarbons and even sold them on the black market across the EU members who had ratified the agreement. The US production and consumption of these products were still rampant due to poor enforcement mechanism and these chemicals continued to contribute to the anthropogenic global warming with the largest recorded hole in the ozone layer recorded in 2006 in the Antarctic. It is important to note that though the process was slow we can celebrate significant progress because 25 years after signing the world had essentially collectively phased out 98% of ozone-depleting substances showing a global commitment but even the newest member state on the planet South Sudan ratified in 2013 (Lazarus, 2008).

Environmental and natural resources law like these shows their importance in bringing about unity and fostering global efforts that go to Worlds protecting both human health and safety together with the environment in which we exist. Without a Montreal protocol it can be agreed upon that ozone depletion but have continued exponentially with the greenhouse effect being multiplied and average global temperatures rising so high that many biomes across the planet would be threatened as well as ourselves from the lack of protection the ozone layer gives us against UV radiation from the sun.

3.3. ENVIRONMENTAL ASSESSMENT MANDATES UNDER ENVIRONMENTAL LAW

Another important form or type of environmental law is environmental assessment mandates which perform three functions:

- They identify a threshold from which potential Environmental ramifications of a specific action will require preparation for;
- They establish a set of goals from which assessments can be endorsed and legally mandated; and
- They sat requirements that ensure assessments will determine whether to continue with the course of action or consider an alternative.

In Legal terms a mandate is a commission by which a body is entrusted to perform a duty without a payment and with security (indemnity) against loss by that body or party. They are essentially environmental Impact Assessments (EIAs) that look at the environmental consequences of a desired action and in a practical sense projects by individuals or companies required to strategically assess their policies and the ramifications they have on the environment. These types of environmental law are a new phenomenon with the increasing environmental awareness that has come within the 20th century. EIAs basically provide the public with a lot of information concerning environmental consequences of a planned course of action and therefore in a way protect the environment before damages even done unlike command and control regulations that directly limit discharges of pollution into the environment (McAllister, 2008).

3.3.1. Environmental Assessment Mandates in Australia

In Australia EIAs and environmental assessment mandates came with the enactment of the US national environment policy act of 1970 and it was first introduced at the state level first rather than the national level. New South Wales was one of the leading states that first created the state pollution control commission and they eventually issued EIA guidelines in 1974. Within the course of that year several acts on both on state and federal level were issued and currently the environment protection impact of proposals act is the central piece of EIA in Australia at the national commonwealth federal level. The state and federal complications in terms of overlap are dealt with through bilateral agreements or the federal government accredits the state and their processes through the provided guidelines for example the environment protection and biodiversity conservation act of 1999 (McAllister, 2008).

Different states have different acts that they use but we will continue to focus on New South Wales specifically. The New South Wales state have the environment planning and assessment act of 1979 that has two different ways for EIAs the first provides state significant infrastructure projects and their impact assessments while the second deals with development assessments for local regional and state Project which is separate from state significant infrastructure.

The federal-level has a legal framework provided by the environment protection and biodiversity conservation act that is enacted also on a local level four important flora fauna ecological Communities and heritage places that define 9 matters of national environmental significance:

- Word Heritage properties;

- National heritage places;
- Wetlands of international importance;
- Listed threatened species and ecological communities;
- Migratory species protected under international agreements;
- Commonwealth maritime areas the Great Barrier Reef Marine Park; and
- Nuclear project water resources in conjunction with Coal Mining development.

Figure 3.8: Australia's great barrier reef protected under the environment protection and biodiversity Act.

Source: https://therightsofnature.org/finally-being-heard-great-barrier-reef-ase/.

When a company or organization (legally called a proponent) is planning a project they have to be assessed for environmental impact under the EPBA act by referring it (the project) to the Department of environment and energy. This information is then released to the public and the state to review whether the project is bound to leave a negative impression on the environment. The Department of the environment and energy then makes recommendations after assessment for the projects feasibility and the final call is down to the minister (Banks, 2018) (Figure 3.8).

One can clearly see why these environmental laws are important as they potentially improve the decision-making of government officials and augments community involvement in initiating and implementing environmental policies.

Environmental Assessment Case Study: (Developed and Developing Countries)

- Germany

Since 1994, environmental protection has been described as a national goal of the German Constitution. It stipulates that the legislature, the government and all public institutions should strive to protect the environment and nature's resources for the profit of future generations. All constitutions in Germany's 16 states also reflect this goal. Many actions of the European Union have influence on shaping Germany's environmental law system. Although different measures have been taken, uniform environmental regulations have not been adopted. Federal and state bodies have various powers in enforcement.

German regulators have been actively formulating policies, as evidenced by the preliminary offshore consent plan. Generally speaking, public awareness of environmental issues is high in the country. From a practitioner's perspective, Germany relies heavily on its permits, warrants and audit systems. In the consent stage, the opinions of experts are mainly used to inform the administrative supervision of industries and operations which may affect the environment. The efficiency of environmental protection usually depends on the related permit system, and in particular whether the facility is required to be audited. (Pinsent Masons, 2018) Accessible at: https://www.pinsentmasons.com/out-law/guides/environmental-law-in-germany

- Brazil

Brazil's EIA legislation was first effected in 1986, and practice has been developing since then. Environmental impact research have been provided to environmental agencies and have the right to obtain permits and perform EIA procedures. Among the most significant controls in Brazilian legislation is the administration control imposed by these institutions. Progress is notable, but implementation among countries is still very uneven. In Brazil, a unique function of EIA is that the prosecutor's office plays an external control role. Proving to society the value of EIA, that is, it conveys real (ideally measurable) advantages, is a challenge that is emerging in the Brazilian community.

In June of 1983, a federal legislation established general provisions in accordance with Brazilian law. According to the decree, environmental permits (links not explicitly established in legislation) must carry out an

EIA. This task is mainly attributed to the states, and the federal government takes action simply on a supplementary context. Therefore, each state must create its unique specialized department or develop a new department within an existing organism, like the health department, to assess environmental permit applications. These specialized departments are called "environmental organs" (hereinafter referred to as "environmental agencies"), and have the power to approve new projects and expand existing processes, and override the authority of other governments.

- Canada

The 2012 Canada Environmental Assessment Act (CEAA 2012) and its legislations have laid the legal foundation for federal environmental assessment practices in most parts of Canada. The objective of CEAA 2012 is to protect the environmental components within the federal legislature from considerable adverse environmental impacts caused by earmarked projects; to ensure careful consideration and execution of designated projects to prevent significant adverse impact on the environment; and encourage cooperation and coordination among federal and provincial administrations.

Brazilian federal agencies with expertise are needed to provide details and advice to support competent authorities in conducting environmental assessments. For federal land programs that aren't designated, before federal authorities or airport authorities can execute the project or practiceany power or carry out any duties or functions that allow the program to be carried out, it should first be regarded that the application of the project won't affect the environment.

- Angola

In Angola, in accordance with the Constitution, the Framework Environment Law (Law No. 5/98 of June 19, 2015) stipulates the basic tenets of maintaining and safeguarding the environment, improving life quality and rationally using natural resources. The key legislation related to environmental matters affecting construction projects and the promotion of sustainable development are as follows: "Environmental Impact Assessment Act" (Law No. 51/04 of July 23, 2004, administered by No. 241/16 of May 25, 2016 Decree Amendment), which stipulates the rules and guidelines associated with EIAs of both public and private programs.

Additionally, the Angolan Environmental Licensing Law (Law No. 59/07 of July 13, 2007) regulates the environmental licensing of activities. The nature, location or scale of these activities may have a significant impact on the environment and society. As well as the "Responsibility Regulations" and "Environmental Damage Compensation Law" (Presidential Decree Number. 194/11 of July 7, 2011). Similarly, there are also some related regulations concerning waste management, water pollution management, and environmental damage compensation liabilities for oil activities and environmental damage.

- Somaliland

Somaliland implemented Forestry and Wildlife Protection Law No. 69/2015, which was signed by the President and published in the Official Gazette on February 2, 2016. This law is the first comprehensive law passed on environmental protection. Somaliland has replaced the 1969 law where the provisions of the law are now inconsistent with the new law since 1991. According to UNEP, the country has also signed some important international conventions on the use and management of natural resources, including: The Convention on International Trades in Endangered Species of Wild Flora and Fauna (CITES); The Convention on Conservancy of Migratory Wild Animal Species; The Regional Convention on the Protection of the Environment of Red Sea and Gulf of Aden; and The Protocol on Combating Oil Pollution and Other Harmful Substances in Emergencies Cases. (Somaliland Environmental Laws, 2018) Available at: http://www.somalilandlaw.com/environmental_laws.html

3.4. ECONOMIC INCENTIVES FOR ENVIRONMENTAL LAW

Economic incentives for Environmental regulation give the motivational drive to achieve environmental goals by encouraging behavior using price signals rather than using instruction and command on pollution control levels or methods in which to do so. These economic incentive instruments vary in type and nature. Countries like Denmark and Sweden and the rest of its Nordic counterparts have imposed taxes on carbon dioxide (CO_2) a meeting companies in such a way that the EU have debated whether to implement the same on a regional level for climate change purposes. This is a gradual process and the merits of the economic incentive instruments must be taken into account and entails:

Integrating the estimated costs of Environmental Services and the emissions directly into the cost of goods services and activities.

- Prioritizing environmental issues;
- Increasing the revenues for Environmental expenditure;

- Stimulating innovation and economic efficiency; and
- Incentivizing customers and producers towards a behavioral change in relation to the environment.

A government structures process in using subsidies grants and tax will determine the price structure and economically drive towards a positive environmental impact can be the general definition of economic instruments and its rational is to improve the state of the environment through voluntary participation. Over the years economic incentive instruments have become more and more attractive to governments and environmental policymakers due to its advantages over the traditional command and control legislature and if properly implemented it could drive quicker towards desired levels of pollution cleanup and the overall cost that both the environment and Society pay. Economic incentive instruments can be divided into five types listed Table 3.1.

Table 3.1: Overview of Economic Instruments

Charges/ Taxes	**Subsidies**	**Deposit-refund systems**	**Market Creation**	**Enforcement incentives**
Product taxes and charges	Grants	Reusable items	Emission Trading	Non-compliance fees
Tax differentiation	Soft Loans	Disposals	Market Intervention	Performance bonds
Effluent taxes and charges	Tax Allowances		Liability Insurance	
User fees				
Administrative charges				

Source: https://www2. mst.dk/udgiv/publications/2000/87-7909-568-2/html/ kap03_eng.htm.

Charges and taxes are essentially levy that are placed on products that would potentially damage the environment. Because companies ordinarily keep a record of the goods that they produce or trade this fee is usually directly related already existing tax collection and enforcement procedures. This makes the whole process substantially easier to calculate.

Effluent taxes and charges depend on the quantity and quality of pollutants released and are supposed to be reflective of the costs to society and because this is extremely difficult to access it's done on a more practical scale.

User fees on the charges and taxes act as compensation for Environmental Services like waste disposal or sewage treatment. Beyond that it includes charges that the plants offering the environmental services pay to comply with overall environmental requirements. These plants are typically responsible for calculating and collecting payment.

Administrative taxes can in some way be also considered as user fees but are paid to the authorities that manage and implement environmental regulations.

Subsidies refer to the States or government granting the organization a sum of money to keep its commodities and services law but in relation to environmental economic incentives, they are used to reduce compliance costs two regulations and work in terms of grants tax allowances and soft loans supplemented through a general budget or earmarked revenues.

Deposit and refund economic incentive systems adds surcharges to the costs of products and eventually this cost is refunded when the products packaging is returned to the collection system rather than being disposed of in more conventional fashions. There are reusable items that still have value to the producer we initially made the product and items for disposal would still fall under the deposit refund systems but have no value to the producer and calls for public intervention to promote the system.

Apart from those examples listed above there are enforcement incentives that sort of fit between being command and control administrative laws and economic instruments that are listed above. The most common way this is done is true the payment of non-compliance fees.

Economic incentives under environmental law are important as they create their own revenue stream and the revenue is directed towards environmental purposes in allocating budgets reduction of taxes and earmarking Revenue. The only flaw is the unrealistic expectations and lack of political skill to execute the implementation through the design of the system. It is true that many organizations or firms comply with environmental regulations for fear of sanctions and other negative impact however economic incentives allow companies to minimize the negative impacts of the activities on nature and be rewarded for it.

3.5. SET ASIDE SCHEMES UNDER ENVIRONMENTAL AND NATURAL RESOURCES LAW

Set aside schemes are exactly as the name dictates and it is characterized by the setting aside of Lands and waters in their natural state and leaving them protected for ecological reasons and biodiversity. From a legal perspective this cannot be done on any other scale but the national federal level as the international community cannot impose on the sovereignty of a nation and at lower municipal levels it is more difficult to implement. This is the skin that was first introduced by the European Economic Community in 1988 to deliver environmental benefits after much damage was done ecosystem due to over aggressive agriculture. It is also a kind of incentive scheme that requires farmers to leave a portion of their land bare, natural or set aside from the intensive agricultural production. One place this worked quite well was in MacSharry, Ireland where in 1992 large arable farmers were instructed to set aside 15% of their land and it was later reduced to 10 in 1996. In 2005 the Farmers who had historically set aside land were awarded entitlements equivalent to the area that they had contributed (Banks, 2018).

These concepts are not unheard of even before environmental laws came into existence where husbandry farming otherwise known as convertible husbandry or up-and-down husbandry was used where by the strip of land was left to grow into grass pasture known as Les and remained like that for 10 years or so before being plowed again allowing time for the environment to naturally tend to itself. Summer Fellows worked under the same concept of allowing the ground to rest over summer when it was out of production regenerating moisture and it was typically a practice done in semi-arid regions or when Dryland farming. Because of the fact that land has become a more sort out after commodity the practice went out of fashion but government incentives has brought it back in a different way with the use of set aside schemes (Banks, 2018).

This has been an important law not only in increasing and maintaining sustainability but also has reportedly had a positive effect on the animals that are kept on the farmland. The EU initially had set aside land policies to increase biodiversity however after the increase of demand in the cereals market the set-aside requirement was push down to 0% in 2008 abolished and set aside completely through the CAP health check. It is however important to recognize the importance of set aside environmental law policies in the way they protect the ecology and maintain resource lifestyle for the health of future generations as well as current ones.

3.6. THE IMPORTANCE OF ENVIRONMENTAL LAW IS IMPORTANT?

The design and application of contemporary environmental law has developed in such a way to respond to the awareness that we have of many Environmental issues that impact the globe. It is understandable however that it has developed in an unsystematic and partial pattern and that can be attributed to a variety of reasons but a lot of effort has been put into identifying the guiding applications and key concepts of environmental law from a general perspective. Several publications have been made for example our common future in 1987 published by the world commission on environment and development and the Earth Summit Rio declaration in 1992. Listed below are some of these laws in application and it is not an exhaustive list and generally accepted in some parts while denied in others nonetheless they bring about a good understanding of environmental law across the world.

3.6.1. Sustainable Development

Environmental laws that encourage sustainable development laws are probably the most widely accepted and most standardized laws in terms of their design and implementation. Sustainable Development globally go by a huge number of definitions and has a variety of uses but from the un declaration in Rio in 1992 essentially the idea behind it is that we should monitor the world such that future generations should have the opportunity to be at least as well off as we are if not better. Specifically, the quotes say that "It is development that meets the needs of the present without compromising the ability of future generations to meet their own needs" it has since turned into a paramount driving principle and concept in international environmental discussions with the 1972 United nation conference on human development in Stockholm the 1983 World commission on environment and development in Brundtland and as already stated the Rio declaration in 1992 (Bosselmann, 2016).

It is a perspective take on economic planning that attempts to drive economic growth while protecting and preserving the environment for the future. It is especially when plan for the long-term heavily dependent on the resources that the sustainability development plan specifically Focuses on. Focusing on forests and their preservation for example will mean the long-term sustained yield of wood and Timber that at the same time does not hinder the habitats of Native birds for example and the population. This idea of sustainability and the laws compounding it are important as an unparalleled responsibility is place on each one of us legally towards us being aware that there are forests to protect and there is a method to the processes we use to protect them. Some legal concepts are still heavily criticized especially when looking at resources that are non-renewable in the sense that from that perspective there is no such thing as a sustainable development goal as eventually exportation will lead to exhaustion (Bosselmann, 2016).

It is a concept that is still taught in schools and application education where there is education for sustainable development in the UN, for example, there is a program that encourages changes in knowledge skills and attitudes to enable sustainability for Society and the UN has been very dedicated in that approach declaring 2005 to 2014 as the decade for sustainable development. This approach has fostered a lot of recognition from the international community and support has continued to grow steadily to target the 17 sustainable Development Goals.

These goals are not laws and have a very of wide-base making them difficult to measure. International environmental laws are also difficult to implement because we must recognize that all countries are at different standpoints with different priorities. Example given is with the Chinese and their air pollution laws being poorly implemented due to the economy being prioritized. If the laws are well implemented and the general public is highly aware of environmental issues many businesses or driving forces in the economy will realize and prioritize financial sustainability and environmental sustainability through activities are approved and patronized by the local communities in order to generate profit (Bosselmann, 2016).

Subsequently financial sustainability requires social and environmental sustainability that can be thought of as the social license to operate that again the local community provides. In this way Environmental Protection is based not only on legislation but on Industries and entire Nations adopting regulations and standards towards their own sustainable development. As much as the UN and other International bodies cannot enforce regulations on States federations or Nations it is widely accepted that international cooperation is key for Environmental Protection and so traditional law is very functional alongside self-regulation. (Georgetown Environmental Law Review, 2019). Available at: https://www.law.georgetown.edu/environmental-law-review/blog/214/ Sustainable development is a very important factor in environmental laws as public administrations are able to form regulations and implement environmental policies as well as reconciling interests of different parties in conflicts of urban planning.

Environmental laws are also important in that they emphasize in caution and extensive evaluation before taking steps that may be disastrous in the end. Because environmental law ordinarily functions in complicated areas of high scientific uncertainty and sometimes extensive scientific knowledge is lacking some of the environmental laws we have are a way of eliminating the possibility of harm. Philosophically environmental laws prove themselves salient in that they recognize that the best course of action is delayed until compelling evidence of rewards and risks are extensively outlined and when the risks are too high or threats are impossible to avert then serious irreversible harm can be done to the environment. Laws put in place avert that. Most if not all human activity is compounded with an element of risk and when deciding something to do we balance risk against possible rewards. The reason why we put the idea of lack of extensive scientific knowledge into the laws we make is because of the complexity of the Natural World and the impossibility of predicting the consequences of unplanned undertakings which could potentially be catastrophic and impossible to repair. The best way environmental laws protect is by removing the element of calculation of risk and possible reward and this has been critiqued by many as unscientific and as an obstacle to advancements and progress. Well designed and implemented environmental laws are advantageous in providing and unprecedented rationale in developing a wide range of International treaties along Environmental Protection as well as health trade and food safety (Macrory, 2008).

Environmental laws are paramount in preventing damage to human beings, natural resources and the environment at large in applications concerning and not limited to global warming and abrupt climate change, extinction of species, novel, and potentially harmful products and technologies to human health and the environment. In this way, environmental laws have been instrumental in resource management and sustainability like in the organization of fish stocks that use harvest control rules (HCRs) and has been used in various Fisheries initially proposed by the International Council for the exploration of the sea as shown Figure 3.9.

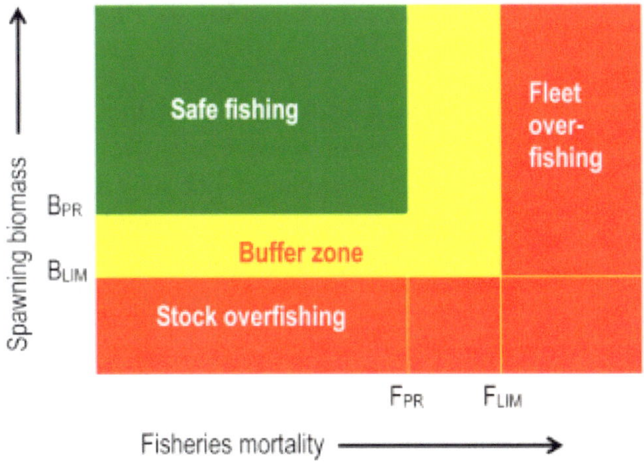

Figure 3.9: The traffic light color convention, showing the concept of harvest control rule (HCR), specifying when a rebuilding plan is mandatory in terms of precautionary and limit reference points for spawning biomass and fishing mortality rate.

Source: https://en.wikipedia.org/wiki/Precautionary_principle#Environment_and_health.

Furthermore, environmental law support sustainability and prevent degradation in dealing with endangered species because of the fact that there is no statistics or scientific information that can be deemed concrete. Environmental laws are vital in that knowing that there is a particular animal or plant that is endangered protective measure will be developed towards that cause. An example given can be that large water body that animals are taken to as a source and the drinking water if contaminated but the exact science is not yet certified then as a precautionary measure the animals are removed from that environment until proof is ascertained and that ensures that contamination will not happen. Another example that has gained popularity in recent times is with the discussion of animal sentience. The question of whether animals are able to experience subjectively the vast range of emotions like humans do is still inconclusive. Due to this fact many philosophers like Jonathan Burke have argued that it is legally ethical to give animals the benefit of the doubt when designing legislations towards the Protection of Animals and environmental laws have gone towards this cause like in the animal rights field (Rostron, 2013).

3.6.2. Environmental Laws Are Crucial in Preventing Cruelty towards Animals

Some environmental laws have been criticized in the way they block of innovations and general progress. There are many advancements especially technologically that bring great advantages but may be banned because of legislature made using the precautionary principle because of their potentially negative impact leaving their benefits untapped. This can be argued for in that for example banning the use of nuclear power plants because of the concerns about the risks related leave us with only the current technology of fossil fuels that will continue to further damage the ecosystem by the generation of GHGs. Philosophically this makes sense however the ingenuity of human beings will never fail meaning that extensive scientific knowledge on the subject matter will ensure knowledge of risks is well understood and therefore actions forward can be taken.

Environmental laws are also pertinent in that a state has authorization on preventing damage within its own jurisdiction and design pollution prevention systems towards things like avoiding discharge of toxic substances in quantities or concentrations that are harmful to the environment and this in concept and practice means early action should be taken to reduce pollution rather than waiting to restore damage to the areas. In principle environmental laws allows action to be taken early stage to conserve the environment and some go further than just legally ensuring that damage to the environment is rectified after they have occurred but preventing them in the first place. This makes environmental legalities powerful and far reaching (Banks, 2018).

The Stockholm declaration of 1972 under principle 21 that states that a nation has a sovereign right to exploit any of the resources in the direction of achieving the economic and developmental goals but it is also their responsibility to ensure that these activities happen within their own jurisdiction and do not cause harm to their neighboring States or anywhere beyond their national jurisdiction. In the Rio declaration raised the importance of sovereignty and as much as it is controversial it is the core element in many conventions and it is imposed and implemented in legislations like the industrial emissions directive in the EU. As stated before, a directive is an official and authoritative instruction rather that voluntarily ratified by states. The criticism with some environmental laws here is that some mandates and even some directives can only be applied by state according to their capabilities and therefore not uniform in any way.

Environmental laws can be developed with an underlying guiding notion behind them that dictate regulation on The Generation transportation treatments forage and disposal of hazardous waste and an example in application is in the Foundation of the Basel Convention on the control of trans-boundary movement of hazardous waste and the disposal. Another example of how environmental laws are important in controlling spaces in which we operate is in Germany where it is stated that no detrimental environmental effects or any variety of adverse effects to the public or environment should be caused and serves to be important in mitigating trans boundary pollution (on a very local level) and in essence preventing pollution as it comes from the source. In this way it minimizes environmental damage and reduces any risks. The underlying point is the preventative approach attempts in anticipating any potential negative effects and uses legal Arms to avoid the damage that will occur. This can be easy to implement as there are various instruments that can be used towards that cause namely:

- Environmental impact assessments;
- Emission standards;
- Best available techniques;
- Environmental quality standards;
- Authorization of hazardous activities;
- Information participation and access to justice;
- Economic instruments for example liability laws; and
- Criminal law.

One concept we have raised through this chapter but not addressed fully is the concept of transboundary responsibility. Environmental laws have been important in fostering the idea that we should not pollute other people's environments as well as our own. We have discussed the sovereignty of states however environmental laws are also designed so that everyone recognizes a responsibility or an obligation to protect one's own environment and to prevent damages to neighboring States. These concepts main limitation is to the sovereignty of a nation. This brings in the economic concept of externalities that is the cost that equates to a third party they did not have any say in incurring the cost and transboundary economic laws limit this idea as it is supposed on other people and their environment.

These integrated Environmental considerations take place not only at the national level but within even smaller jurisdictions like companies themselves taking up charge in the decision-making process through EIA mandates and other provisions. Transboundary and integration environmental laws are important and necessary towards cohesive and coherent policies for sustainability and that their major importance (Banks, 2018).

Environmental laws also require public participation and transparency and this is important for accountability of government and industrial concerns. This solidifies the idea that for effective protection of the environment and human beings it is the public's rights (that is essentially all of us; you and me) to hold and express opinions on ideas related to information given by government and Industries on their Economic and Social policies regarding the environment around them. Legally, human beings have a right to access the information regarding actions taken towards environmental policies.

International environmental laws outline objectives that inspire other specific national environmental laws that will eventually end up being directives, mandates or voluntary instruction. Inclusivity that is involved in environmental law also provides a reference to carry the community and other institutions either local or International to adhere to norms that at the end of the day serve as direct positive impacts to the environment. Some laws are also set with autonomous normative principles for developmental activities in making criteria to evaluate homogeneity and this is subject to the activities in question together with the parties involved and ranges in practicability and effectiveness. There is critics that say that some environmental laws have proliferated in such a way that they have become ineffective due to the fact that if there is such a wide and diverse range of environmental issues economic and political considerations to take into account. The balance of interests between socioeconomic factors and becomes very difficult and sensitive when taken into account (Lazarus, 2008).

Environmental laws also hold important facets within them that hold parties responsible for producing or disposing of pollution. By definition parties may they be a person or a group of people that compose a single entity are liable for the damage done to the natural environment and it is a policy that has garnered strong support within the organization for economic cooperation and development where it was first recommended in 1972 and further reaffirmed in 1974. It is a fundamental concept in environmental law in the United States of America and EU members creating that idea that the one responsible for pollution pays for the damage that they caused.

Allocating costs of pollution prevention and designing control measures to promote efficient use of dwindling resources and to avoid imperfections in international trade and investments and holding the polluter responsible to bear the expenses of carrying out measures to remediate the negative effects that they have on the environment is useful in raising both legal and environmental awareness. These costs are decided by public authorities within the state. In this sense the laws are designed to cover liability in cleanup costs due to damage of the environment. Defining pollution, prevention, and control measures and their effects is one practical way mitigating pollution at the source (Lazarus, 2008).

Environmental laws in this way hold substance when discussing what is the pollution and in this way if pollution exist what administrative thresholds are exceeded and this quantifies to unlawful Act. It may not be an unlawful act if the pollution is defined independently by administrative thresholds. This is interesting because the pollution that is self-emitted by the polluter does not necessarily mean it has already affected the environment negatively but minimization of risk has to be paid by the potential polluter and in this way it overarches into the preventative and precautionary principles of environmental law.

Another perspective element in substantiating many environmental laws is by questioning who the polluter is and a polluter in practical reference is defined as an activity that is harmful towards the environment but in legislation it is in a much more extensive fashion. This means that those who have not actually polluted the environment but any party whose actions is causing risk for the environment is liable and counts as a polluter again showing the importance of environmental awareness in respect to the law. A great example of this is the Erika oil spill case where the European Court of Justice found the producer of hydrocarbons liable for cleanup costs for contaminants that were well out of their hands. This is because the risk of pollution existed even before the product was boarded and the producer is meant to be aware on what impact the product has.

Another factor concerns how much is to be paid by the party that is liable and this is dependent on three factors; the cost of preventative and precautionary measures the administrative procedures and the cost for reinstatement. The amount that the parties liable are required to pay or Limited by the principle of proportionality because only the cost for own pollution is meant to be paid. These are some exceptions especially when joint and several liabilities are the cases (Macrory, 2008).

Different instruments used under the different facets of environmental law to design manage and implement environmental laws in this way include:

- Licensing procedures;

- Prohibitions emission limit values;
- Subsidies and feed in tariffs;
- Certificates, tax alleviations;
- Liability rules, voluntary agreements under soft laws;
- Environmental management systems;
- Labeling; and
- Orders and sanctions.

Different types of environmental laws and their importance as explained above is no exhaustive list but represent some important postulations in the functioning of environmental law locally and globally. They are important in making sure that the law is efficient and effective in a practical pragmatic way. It is timely however to realize the recent confluence of national and international issues on energy, economics, and the environment as a highlight to the complications and importance of environmental policy, law, and decision making. The complication here arises because a decision in one element, may it be economic or political will overlap into the environmental sphere and vice versa.

3.7. EVALUATION OF THE SOCIO-ECONOMIC BENEFITS OF ENVIRONMENTAL LAWS

Environmental law has proved itself beneficial and salient from Social and Economic standpoints as well as Environmental protection and Conservation. Qualitatively types of benefits and importance of environmental law have proven themselves through all facets of life. From a human standpoint environmental laws have helped protect the air water and soil together with all other medias of the environment towards positive health benefits.

With the air environmental laws have been cardinal in deterring respiratory illnesses and premature deaths because of improved air quality. With the water the health benefits have involved getting access to cleaner drinking water that isn't polluted and with waste there has been a reduced risk of contamination poisoning and accidents due to leakages from landfills and other waste disposal facilities.

The Resource benefits of improved air quality due to environmental laws have made sure there is decreased damage to buildings and crops raising productivity. Resource benefits from cleaner water due to environmental legislation include having pristine water for companies to work with and in waste management it has meant that environmental laws are important in reduced-input of primary material and more conscious processes of production.

The types of benefits that environmental law has had on ecosystem if you would look at medias that are specific environmental attributes include the air having been cleaner by declining global warming through decreasing CO_2 emissions and banning hydrofluorocarbons and chlorofluorocarbons. Water quality has increased in ecosystems improving river and lake quality although there needs to be plenty of work done on the ocean more specifically the great Pacific garbage patch that is an Unbelievable 1.6 million square kilometers of offshore plastic accumulation. In relation to waste disposal management and the environmental laws that have brought forward in recent times have been beneficial and important and decreasing methane emissions from landfills and by extent we have avoided another facet of global warming.

When revealing the importance and benefits of Environmental laws we should not only look at the qualitative benefits of Environmental laws but quantitatively what is the extent of these benefits and in relation to value can we actually monetize these benefits? In this case, we will take the effect of EU mandates directives and legislation and relate their importance to the effects that they have had on member states.

When observing the extent of the benefit that environmental laws have had it involves looking at some figures to put matters into perspective. In the EU the directives mandates and legislation put forward have resulted in air emission particulates dropping by an estimated 1.8 to 3.3 million tons and that is a phenomenal step forward. Translated into the health sector it means that between 15000 to 34000 cases of premature deaths across EU member states have been circumvented.

The extent of the effect of legislation has meant that candidate countries in the EU have experienced improved drinking water quality. 20 to 30% of households in Turkey Bulgaria and Estonia that did not have water supply currently do because of the laws that have been put in place. Water quality has also improved in color taste and smell due to reduced contamination.

In regards to waste methane emissions were the biggest concern and through the landfill directive we have seen a drop of up to 6.4 million tones and this is in spite of 2% growth in waste generation. What this means is that legislation has fostered better landfill waste management and improved recycling.

Looking at nature and the environment with Christine European ecological systems the size of protected areas in Europe has and continues to increase. In countries like Malta they have increased protected areas by 10 percentage points from 18% to 20% for example.

Value benefits of designing implementing and enforcing environmental directives have brought about economic values which for many is most likely the most important function of Environmental laws. Implementation of EU directives between 1999 and 2020 will have accrued interest of between 134 to 681 billion Euros assuming full implementation is achieved. Countries that will benefit the most include Poland Turkey Croatia and Romania and are listed in Figure 3.10.

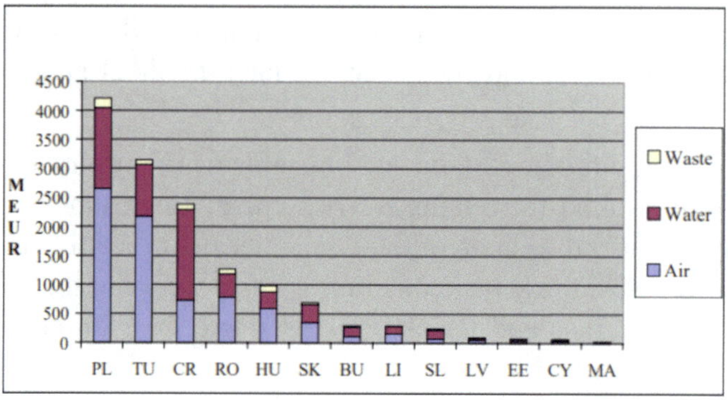

Figure 3.10: The annual value of benefits for full compliance to European Union directives in millions of Euros.

Source: https://ec.europa.eu/environment/archives/enlarg/pdf/benefit_short.pdf.

In summary, we can conclusively state that laws are put in place for a reason and they are designed either for the public well in protecting and keeping them healthy and others are for the sake of avoiding any form of danger. Good environmental laws are able to balance between financial issues and future-proofing our world for future generations to enjoy. The main purpose of environmental law may be to preserve the environment but it goes beyond that in ensuring things like food and water security for everyone. Better yet Environmental legislation does not actually cost money like people assume but in the long run potentially mitigate this course by creating new economic and revenue streams 10 times over what would have been spent on curative measures like expenditure on Health

Care. Derivative effects of environmental degradation goes as far as forcing political instability and global insecurity which means that the abstract importance of Environmental legislation should be made more transparent in order to foster global unity. In fact if anything it has been proven that although in the short term, implementation of new environmental laws may produce a short term loss in economic progress in the long run the benefits are very noticeable.

Creating and educating over environmental awareness are a way to make sure that human beings do not further degrade the environment and maintain it for current and future generations however it has been proven that legislation is the only way that we can ensure that environmental conservation is actually practiced in everyday life. The laws around Environmental conservation make sure the company's businesses and the general public behaves in a way that both prevents pollution and preserves the environment from degradation by laying down strict punishment for those who fail to obey the rules. It is this kind of enforcement that guarantees philosophies ideas and schemes are translated into real effort in conservation spirit online International level there is a court in The Hague but there is no way it can actually enforce decisions and Nations I have in the past ignored Court orders made by them that are not in their interest or in their favor but we cannot deny that there has been great progress made towards environmental protection on a worldwide level.

The use of environmental law also requires some sort of decision-making that is integrated where we incorporate environmental social and economic considerations towards the goals governments realize that environmental health safety regulation industrial trade and employment policies or need to be considered in order to make incremental improvements integration also Fosters technological innovation because their mutually reinforcing objectives that represent a large number of options for achieving environmental goals. This works especially with sustainability where many observers will recognize it requires new laws and modification of existing laws.

Laws exist for a reason and they are statutes that bind all people living in a community and the world at large. Laws protect our health and safety, and ensure our rights as citizens against abuses by other people together with the rights of animals and non-human entities, by organizations (with their internal legislations), and by the government itself. We have laws to help provide for our overall safety. They should be governed by accepted rules by the greater public, rather than by the arbitrary decisions of those in charge. All law is important, because if we do not follow the rule of law, we have chaos and this applies with environmental law as in that if we do not have them we will suffer from environmental chaos.

4

Water Management

CONTENTS

4.1. INTRODUCTION

Water! Well, what is water? It is said that water is life. Does this mean that we cannot survive without water? This might be true. Let us examine our environments. (Sargen, 2019). Available at: http://sitn.hms.harvard.edu/uncategorized/2019/biological-roles-of-water-why-is-water-necessary-for-life/ We are always surrounded by water and water is surrounded by us. Our board rooms are ever filled with water bottles, our classrooms, our offices actually do have a dispenser, our houses, name it all, there is water everywhere. We also carry water with every place we go. Our habits do indeed prove the statement that water is life to be true. For life does not really mean being able to breath, it means what we treasure most and pushes us to continue with our never ending growth in this globe, and water is one of them. The entire globe is also said to be composed of more water than dry land, hence the phrase that we are surrounded by water and water by us for the small available dry land is filled with us human beings.

In simpler terms, water is basically a fluid whose more of a product of a chemical reaction between two elements which are hydrogen and oxygen. It has no specific color and the taste of water cannot be really defined. With water, one is able to slightly see through it more so if it is enclosed in a transparent container. Water has no specific smell, however, dirty water is known to have a specific smell depending on the type of pollutant used.

Water is known to be obtained mostly through rain. Atmospheric moisture does rise to a certain level known as the condensation level in which the rain bearing clouds are formed hence occurrence of rain. Water can also be obtained as a result of the melting of snow and ice, which is the solid-state of water. Ice is mostly found in the extremely cold regions of the globe while snow is found in high latitudinal levels more so mountains which are above a certain latitudinal height.

4.2. CHEMICAL AND PHYSICAL PROPERTIES OF WATER

As stated earlier, water is a product of a chemical reaction between two chemical elements. It involves the reaction of two hydrogen elements and one oxygen element for it to be completely formed. Water is also known to be a universal solvent for it can dissolve multiple solutes to form a uniform mixture. This means that before water is used let's say for domestic purposes, it must be cleaned for naturally water contains several dissolved components in it.

Water is a unique chemical which can exist in three distinct states. It can exist as either a liquid under an ambient temperature. This is a temperature that is not very high and not very low. Under high temperatures, water can be heated to be in a gaseous state mostly referred to as water vapor or steam. Water vapor is basically tiny droplets of water which are not clearly invisible in the air. The other state of water is formed once water under normal temperature is exposed to extremely low temperatures making it freeze hence formation of ice. The three states of water can be interchange when exposed to different processes. The process of converting normal water to steam is known as evaporation. The process of converting water to ice is known as freezing. Ice can be converted directly to steam through a process known as sublimation. Steam or water vapor can be converted directly into ice through a process known as deposition. Water vapor can be converted into its original state of being liquid water through a process known as condensation.

The rate or mass per unit volume of water, commonly referred to as density does differ since water has different states. Water density reduces as water changes its state from liquid form to solid form and it increases as water changes its state from liquid form to gaseous form. The density of ice is way below the density of water in its ambient temperature. Such a characteristic is of great advantage to aquatic life. This is because, in nations which experience long periods of winter, freezing does affect their water bodies. Since ice is lighter than water, it floats on the water bodies insulating them from the cold temperatures in the surrounding hence maintaining their temperature at a certain level in which aquatic life can still thrive normally. The destruction of aquatic life is prevented due to the density differences between the various states of water. However, the ice could be of great danger more so to navigators. This is because the visible part of the iceberg seems to be small from the top yet its submerged volume is quite huge causing water accidents to navigators who misjudge their size.

The melting and boiling points of the different states of water do defer depending the pressure level of their surroundings. Under high pressure, ice is likely to melt at higher temperature and in an increased rate as compared to when it is under low pressure. The boiling point is also affected in that, at high pressure, the boiling temperatures are reduced and at low pressure the boiling temperatures have to be increased for the desired results to be obtained. All this pressure issues affect cooking in both high land regions and lowland regions. This means that people residing in highland regions which end to experience low atmospheric pressure have to take longer preparing their meals or boiling water for the temperatures must be increased for the desired results to be obtained. As for the lowland areas, the atmospheric pressure is quite high meaning their temperatures will hit their maximum level within a shorter duration for their pressure is high.

The states of water do have two points in which they can interact. We have the triple points which is basically a certain temperature in which all the three states of water can coexist without any effect. The second point is the critical point. The critical point is the temperature in which the three states can no longer coexist together.

As stated earlier, water is known to have no specific taste or smell. However, living organisms such as frogs are said to be able to smell water from a far. The sensory system in human beings does have the ability of detecting the presence or let's say the taste of water in their mouth. This does not mean water has a specific taste because, despite humans being able to detect presence of water in their mouth, they cannot state a specific taste or detect a specific smell. Water is quite a good solvent hence it contains several dissolved substances which might give it a specific taste or a specific smell.

Water is naturally colorless. However, water bodies with grown algae deep in their water beds, reflect their waters to be green in color in the presence of sunlight. Pure water in a clean glass placed against a white surface reflects its color to be blue for water does reflect the color of the skies which are blue. With no reflection or contaminations, water is generally colorless.

Water is quite a good polar substance for it does dissolve various substances to form a uniform solution. Gases do also dissolve in water. For example, carbon IV oxide does dissolve in water leading to the formation of a uniform fizzling solution. However, not all substances are soluble in water though a good number does defy the odds of insolubility.

Availability of two hydrogen atoms in water increases the strength of its bonds. This means that water does maintain its particles together with the help of a very strong bond that makes it hard to tear apart the three atoms of water. Such a property increases the surface tension of water for light materials rarely sink in water unless the water is contaminated. The capillarity action of water is enhanced for the water molecules will be strongly attracted to each other hence their ability to rise up a narrow tube with minimum distraction.

Water is not such a good conductor of electricity. However, with the increase in the density of water by adding contaminants such as salt, the electrical conductivity of water is increased. We did mention that the water atoms are strongly held together by strong forces, however with the passing of very strong electric current through water, the forces holding the water particles tightly together are weakened leading to the separation of the water atoms.

Water is said to be incompressible. This property gives water the ability of being a good medium for sound transmission since the chances of decrease in the speed of sound transmission along the pathway are limited.

Highly electropositive chemical elements are said to be able to displace the hydrogen atom from water destroying its final structure.

Water is said to have various phases in its different states. The phases of ice which is the solid water do vary with the temperature. The phase of liquid water does vary with different degrees of the ambient temperature. The phase of steam or water vapor, water in gaseous form, do also vary with a slight change in temperatures.

4.3. EFFECTS OF WATER ON THE EARTH'S SURFACE

As stated earlier, water covers almost two-thirds of the earth's surface. It is all around us in rivers, lakes, dams, water ponds, oceans among others. The sea and ocean water is more on the earth's surface as compared to the other water bodies. Water exists in solid form in cold regions, in liquid form and vapor forms with some water being termed as underground water in aquifers. Water does also affect the underground activities from the mantle for water is part of the mantle which leads to the volcanic activities on the earth's surface.

4.4. THE WATER CYCLE

Water Cycle refers to the movement of water from the underground, to the soil to plants, to surface water and the atmosphere.

The groundwater is absorbed by plants for their growth and to help in transportation of plant nutrients from one part of the plant to the other. On the surface, plants transpire the excess water from the system into the atmosphere. Animals do also get rid of the excess water in their system through sweating among other ways. The sweat droplets are released into the atmosphere. Water from bodies such as rivers and oceans evaporates to the atmosphere during the day to the high temperatures from the sun during the day. All the moisture accumulated in the atmosphere continues to rise during the day due to the low pressure in the atmosphere up to a certain level referred to as the condensation level. At the condensation level, rain bearing clouds are formed from this moisture and this leads to rains. The accumulated moisture does also lead to the formation of few in dawn times when the moisture comes into contact with the cold morning temperatures. All this forms of precipitation lead to the release of water back to the earth's surface. Runoff water gets infiltrated back to the ground, some sips into water bodies, some is used by animals for consumption purposes and the rest used to water plants. The water cycle is a basically a never-ending movement of water from the ground, to the atmosphere and back to the ground.

Lack of balance in the water cycle might cause serious negative issues. It might lead to flooding whereby the precipitation rates are very high for the water to be accumulated back to its original sources. It might also lead to drought whereby the water supply does not meet the water demand in the region.

4.5. WATER STORAGE

Most freshwater is said to be stored in the solid-state. This means that most freshwater is either in snow form or ice form and can only be sourced from cold regions and high mountains which are above a certain altitudinal level. Freshwater is also available in lakes and ponds, however, the volumetric capacity of lake water does vary depending on the season. In wet seasons, lakes, and rivers tend to be full due to the high rains available while in dry seasons, the water level in lakes and rivers retreats back.

Salty water covers the largest percentage of the earth's surface. This is because it is contained in both seas and oceans and the underground water. Underground water is salty due its contact with the underground rocks which contain chemical components. The seas and oceans are salty because during their formation, the sea beds were accumulated with volcanic materials. Volcanic activities do continue to take place on the sea beds increasing their salinity. The sea waters are affects by tides which are caused by the pulling effects of the Sun and the moon in relative to the changing positions of the earth due its rotation. The tides help in the determination of the current water level for they leave behind a tidal zone. In situations where the tidal zone decreases in size, the seawater levels are said to have decreased in size while when the tidal zone increases, the water level are said to have increased.

4.6. IMPORTANCE OF WATER TO LIFE

We did say water is life. Water helps in growth of all forms of life. Animals depend on water for various purposes such as metabolic processes. Metabolism does use water for the replication of various body nutrients. Enzymes separate both the hydrogen atoms and oxygen atom from water. The distinct atoms are used in the formation of substances which provide an animal's body with the required energy for its survival.

Plants need water to survive. Water helps in the transportation of manufactured plant food from the leaves to the other parts of the plant for growth purposes. Absorbed nutrients are also distributed evenly through the plant structure by water. Plants absorb sunlight energy which aids in the splitting of the hydrogen water atoms from oxygen atoms. The hydrogen atoms aid in the formation of glucose through its oxidation as the oxygen atoms gets released in the atmosphere for human race to breath in clean air.

Water does also help in neutralizing different acidic and basic components in the human body for water is said to be neutral in nature since it has a pH level of seven.

Water does support aquatic life. Aquatic plants such as algae among others depend on the air circulation in water bodies for their survival. Water does also allow penetration of sunlight into the deep parts of our water bodies hence supporting photosynthetic processes of aquatic plants. The formed food by aquatic plants is fed on by aquatic animals. Fish, dolphins, whales among other aquatic animals do depend on water got their survival. This is because they can barely survive on fry land for long periods. Fish use gills for breathing in water. Dolphins and whales constantly come out of water to breathe in fresh and go back into water for breeding among other survival activities.

4.7. IMPORTANCE OF WATER ON CIVILIZATION

Most civilized regions around the globe are said to be close to water bodies. Regions such as coastal regions and areas close to water bodies do owe their success to availability of water which has helped in invention of economic activities such as navigation businesses, fishing, agriculture among activities.

Water is mostly used for domestic purposes. By domestic purposes I mean that it is used for preparing meals, washing clothes, taking a bath among other activities. Domestic water after use ends up being dirty water. Since clean water is a scarce resource, the domestic water is treated for reuse instead of being disposed off and polluting the environment.

Agricultural activities are the main supplies s of food around the globe. Most countries do also depend on agriculture for their economic growth. In the past, most farming activities relied on natural precipitation for the survival of farm produce. However, with the increase in human population which lead to increase in the demand for food, farm produce has to be produced in huge amounts. This demand has brought about the invention of various irrigation techniques so as to ensure there is constant supply of food in all seasons, be it dry seasons or wet seasons so to meet the high demand. Availability of water ensures that the irrigation systems never dry up and enough food is produced in our farms.

Fishing is part of the agricultural practices being carried out by farmers. Farmers have opted for the inland fishing techniques whereby fishponds are being constructed away from rivers and filled with fish fillets. The fillets are well taken care of and later sold in the market for human consumption or even for medicinal purposes. The fish pond idea ensures that we balance out biodiversity for there will not be overfishing practices in lakes, rivers, and oceans.

Water is largely used for drinking purposes to all animals. For example, the human body is mainly made up of water more than it is made of blood. Human beings and other animals require water to cool their bodies in hot seasons. Apart from cooling purposes, water is required to help in the prevention of constipation for with enough water, the human digestion process is said to be conducted in the most effective way. Water does also help in the cleaning of waste from an animals' body which is later released out of the body either in form of sweat or urine.

For ages, human beings have relied on water bodies for navigation purposes. With increased connection among different human races around the globe, water transportation has become a major navigation means. This is because trade activities have increased and there is need for trade goods to be transportation from one place to another around the globe. Sea travel activities are no longer a thing of the past for our current generation is opting for the travel around the world through water systems as they get to know their globe better.

Water is of great use in conduction of chemical reaction. It can be used as a catalyst, in some reactions so as to speed up the rate of reaction of the available reactants. It can be used as a reactant in manufacture of certain chemical products. The main reason it is largely used in chemical reaction is that water is a polar solvent which is able to dissolve various solutes and fellow solvents. Water is amphoteric, that is, it is both acidic and basic hence not largely affected by either of the acidic or basic reactions, in fact, it contributes a great deal to them.

Human civilization has led to the establishment of industries almost everywhere around the globe. The industries are known to manufacture different products. The manufacturing process does end up producing a lot of heat through the machines used to produce products. For effective functioning of the machines, they need to be cooled regularly. Water channels are incorporated in the machines during their construction to ensure that there is proper supply of water which aids in the cooling process. Nuclear power plants do also use water as a coolant in the reactions taking place in the core. Water can also be used as a moderators in the nuclear core as it tries to moderate the speed at which the core reactions take place by either increasing the reaction rates or reducing the reactions rates or even maintaining them at a constant level.

Water can be said to be a good fire extinguishing component. This is because once water comes into contact with heat, it vaporizes. The vaporization of water takes away the heat from the fire slowing the burning effects of the fire. However, water is not highly recommended to be used as a fire extinguisher more so if the fire has been caused by organic substances such as coal, charcoal, and petroleum among others. This is because such organic materials react instantly with water and instead of slowing down the burning effects, water assists them in spreading the fire to a larger region or it can increase the burning rates.

It is said too much with no play makes someone a dull person. Well, water is always present to contribute in the fun part for human beings and try and reduce their workload stress. Water is used for recreational purposes. Beaches are known to be tourist sites in many places for it is believed that the sight of flowing water and the sound of water generally does help in calming us down. Swimming and water skiing are fun activity to humans. With ice being a state of water, it is also used for recreational activities such as ice skating and others. Sport have been introduced to help in entertaining people and also as a livelihood for the sport people. Most sports played on or in water are the likes of ice hockey, ice skating, sledding, snowmobiling, snowboarding among others. Manmade water fountains are all over the place for they serve as decorations in our homes, offices among other places.

With water being a major part of every living thing's life, water industries have been setup for different water purposes. Some have been setup to help in collection of water. This means that they have systems for harvesting rainwater. Systems for obtaining underground water have been setup. Sewage treatment systems have been put in place to ensure that there is constant supply of water since water is a limited resource. Artificial borings have been dug almost everywhere more so in dry lands to ensure that there is constant supply of water to human beings. Basically, the water industry ensures that there is constant supply of water to all living things hence it does its best to obtain enough water from all the possible sources.

We need energy for us to run different activities. Water is being used in the production of hydroelectric power. This is basically done by using water under very high pressure to turn the turbines which are connected to a generator hence release of electricity. This reduces the pollution caused by the use of other sources of energy. Pressurized water can also be used for water jet cutting and water blasting.

Water can be used for medicinal purposes. Most drugs are manufacture with the help of water more so the liquid drugs. For to take drugs in form of pills, they have to ingest them with the help of drugs.

Food processing activities do make use of water. Boiling meals, steaming meals among other cooking ways of make use of water before the food is said to be ready. Some food flavors such as salt are known to be by-products of water and without salt, most foods are said to be tasteless.

4.8. DISTRIBUTION OF WATER

We did mention that water is almost all over the continent. This is true for water exists in three states. Liquid water is said to cover two-thirds of the earth's surface. The remaining percentage is definitely filled in by the other states of water. This is because high mountains tops are covered with snow, which is a state of water, cold areas are covered with ice, which is a state of water, and the atmosphere is filled up with water vapor totaling the percentage to be a hundred. So it is true to say that, water is part of us and we are part of water as a universe.

Research has it that every planet on the universe does have a certain percentage of water. The difference is that not all planets are able to accumulate water in its three distinct states. This because, for example, a planet like Mercury is way too close to the sun hence ice water cannot exist due to the high temperatures from the sun. The space is said to contain more water than the water said to be in the earth's surface. Basically, the entire universe contains water.

4.9. THE CLEAN WATER ACT (CWA) USA, 1977

Due to the much importance of water to the entire universe, different governments have made it their responsibility to protect it and ensure that it is maintained in good condition always. The government of the United States of America made it their responsibility to come up with a law that ensures that water is generally protected from pollution or any kind of misuse. The law was named the CWA. It insists in the maintenance of water in its original physical, chemical, and biological condition at all times no matter the situation in place (Mittal, 2010).

The contamination of water is known to not only occur at the water source but also in the many distribution channels of water. Contaminants maybe naturally occurring such as uranium materials which available more so in nuclear power plants. Water can be contaminated due to the release of harmful chemical substances to the water streams such as sewage water which is quite contaminated. Human activities such as farming may lead to contamination of water for farming activities involve the use of fertilizers which might be washed away into rivers hence contaminating them.

One of the major reason for the establishment of the CWA was because of the negative effects it was having on the health of human beings and all living organisms in general. Intestinal complications, reproductive issues, neurological infections, pregnancy complications, infant illnesses among other health complication have increased in number due to the use of contaminated water. Individuals with permanent health complications are faced with great danger due to their exposure to contaminated water. This is because their health system is already weakened and further exposure to contaminants will only render their immunity almost useless to them (Mittal, 2010).

Gastrointestinal illnesses such as anal sores, constipation, perianal infections, colitis, infections of both the small and large intestines, esophagus disorders among others are majorly said to be caused by use and the consumption of contaminated water. Such health risks are known to affect majorly the elderly people and the young children. This is because, for the elderly, their immune system has weakened over the years hence no longer able to fight bacteria from contaminated water. The young children have their immune system still growing hence it is not strong enough to fight such bacteria's. The rise in the number gastrointestinal infections has really been a major concern which lead to the immediate implementation of the CWA.

Reproductive illnesses have been part of human beings for decades. However, infertility issues were mostly associated with women and not men. This does mean men are not faced with infertility issues, it's just that the percentage of men affected by such issues is quite minimum. With the increase of water pollution issues around the globe, infertility cases have increased among the men more than in women. Research has it that most contaminated water contains a certain chemical component that alters the functioning of the male hormone know to be a major contributor in the fertility of men (Houck, 2002).

The neurological system of the human race is very delicate. It requires all forms of medical care to prevent any damages to it. It composed of the brain, the spinal code commonly known as the backbone and the nerve system. Consumption of contaminated water for prolonged duration does lead to brain disorders, spinal cord infections and the improper functioning of the nerve system. Main contaminants associated with neurological problems include insecticides, pesticides among other such poisonous chemical. Infants and unborn babies are the ones mostly likely to suffer from neurological disorders since they slowly develop with the water contaminants already in their system (Houck, 2002).

4.9.1. Objectives of the Clean Water Act

CWA was mainly implemented for the following reasons.

One of the reason was that the act was to ensure maximum protection of water on the surface and underground water by clearly stating under which conditions water as a natural resource was to be used more so for the economic purposes. The act does also ensure that there are clearly outlined conditions under which the negative effects caused availability of excess water such as the flooding issues. Negative effects such as drought due to scarcity of water and general proper management of water are clearly taken care of in the CWA.

The second reason for the implementation of the CWA was to ensure that there is a clear difference between surface water and underground and their sources. This was included in the act to enlighten the public on the difference between the two. With this difference, individuals are able to understand the legal laws protecting surface water, underground water and the relationship between the two. This ensures that lands close to surface water are well-governed under certain laws, same to grounds known to supply the public with water from the underground (Houck, 2002).

The last major objective was to ensure different terms to be used in protection of water were clearly defined to avoid some common confusion mistakes during their use.

One of the terms is surface water. Surface water is to be used when referring to water channels naturally available on the earth's surface. Sometimes such water channels maybe disappear underground and reappear at the other end. So long as the water channel does reappear at the other end, it is to be referred to as surface water.

The other term was groundwater which is to be used when referring to water naturally occurring beneath the earth's surface. It is basically not visible from the ground level for it in direct contact with the parent rock of the ground and it is covered by layers of soil above it. Since most drainage systems have been constructed to transport water in dug underground channels, they are referred to as groundwater channels for at least are covered by some few layers of soil (Stephenson, 2005).

Water body is a general term used to define either a surface water channel or underground water channel so long as the channel is known to contain water and also be able to transport water. The channels maybe artificial it natural.

Water resource is used to refer to water channels beneath the ground once they are being used to supply water above the ground surface.

The use of either surface if underground water basically means the different activities in which water from either sources is made use of in different activities such as domestic uses, recreational purposes, industrial purposes among others.

A Hydrological region is a term used to refer to an areas surrounded by various sources of water, be it surface water, be it groundwater, it doesn't matter, so long as the water is naturally occurring.

A river basin is used to describe the areas into which surface water bodies such as rivers drain their water into. It more like a combination of multiple rivers emptying their water in a common region.

4.9.2. Water Rights

Any water source is not subject to ownership to anyone. By this, it means that, no individual can claim ownership over both surface water or understand water. Be it that the water body is running through someone's property, the owner of that property has no right to claim ownership over that water body.

If water has been harvested or drawn from its original sources, be it surface water such as from a river of groundwater such as sourced out from an underground aquifer, it can no longer be said to be part of the surface water or the groundwater (Stephenson, 2005).

In some situations, the government might be conducting a survey on where a certain location receives its water from. There might be confusion on whether the water is from a surface water point or from groundwater storage. The water authority, which is known to handle issues to do with the CWA, is known to be responsible for the determination of where the water is from.

For industrial purposes and other special uses, certain water sources may be declared to be the main suppliers of the amount of water to be used. This decision that a certain water body is to supply water to industries for use as raw material or for other industrial uses, needs to be well documented and presented to the appropriate authorities for recording purposes and for reference in future instances (Houck, 2002).

It is a water right to distinguish between surface water and groundwater. For mining sites, once groundwater has been exposed during excavation of mineral raw materials, it is no longer termed as groundwater but as surface water for it has now been exposed to the surface. Only before being exposed to the surface it is to be referred to as groundwater.

The use of either the surface water or the ground water must be done in a way that ensures the water is well protected from any form of contamination. If the water is being harvested for economical purposes, the legal rules and regulations associated with the use of water for economic reasons should be adhered to by ensuring that it is efficiently used (Houck, 2002).

Manufacturing industries which make use of water either as a raw material or as a heat exchange component, have the role of improving their technology. The technology should be enhanced such that it ensures that only the required amount of water is used as a raw materials or as a coolant substance so as to avoid misuse of water. The technology also does ensure that no water can be released back to either the surface water or groundwater before being properly treated according to the rules and regulations governing the treatment of contaminated water.

Buildings are being constructed with each passing day all over the world. As stated earlier, we as human beings we require water for our survival for we are part of water as it part of us. This means that most of the buildings need to incorporate with water systems during their construction. It the right of water that the water systems being installed in our buildings which are either being renovated or newly constructed is in perfect conditions and known to perform its functions effectively with minimal drawbacks. Since once the water has been used, most of it ends up being contaminated water after use. Drainage systems do need to part of our current buildings. If possible, treatment plants need to be incorporated in the drainage systems to ensure water is treated within the same building and reused again for the same purposes or different purposes. This needs to be done to avoid the misuse of water since water is a limited resource (Mittal, 2010).

The withdrawal of water from either the groundwater or surface water is quite regulated under this act depending on the kind of use it is to be used for. However, if individuals are withdrawing the water for their personal use and not special technical use, they do not really to be certified to withdraw the water.

Individuals are allowed to change the course of the surface water bodies in their surrounding just slightly with the help of water experts in situation where the waters are being of negative effects to them. The construction of private artificial borings is also allowed and no certification is required. Harvesting of rain water does not need to be certified by the government for it part of human rights to ensure that they have constantly supply of water for their domestic use.

Surface waters need to be protected for they are life to many living organisms. It is the right of water that individuals should stop farming in the river banks for the farming activities lead to damage of the water body in one way or the other. Release of waste into rivers and other bodies is strictly prohibited for it leads to contamination of the water which does pose a great danger to the aquatic life. Fish breeding grounds need not be interfered with for this means that most fish species might end up being dead (Mittal, 2010).

We did discuss the use of water for navigation purposes. Different vessels are used for transport both cargo and passengers through various water bodies. Under the act of clean water, no authorization or certification whatsoever is required to allow vessels to travel via the various water bodies. The withdrawal of water either for cleaning purposes or testing purposes in the various water body vessels does not need to be certified by the water authority. It is their right to withdraw such water without any permission from anyone.

Pollution does occur everywhere. Even in the water vessels used for different purposes, there is pollution. It the duty of vessel operators to ensure that the vessel is well equipped with appropriate tools to handle any wastewater or any kind of waste material. The waste materials should be regularly examined to ensure that they are always in good condition to avoid contamination of surface water as the vessels travel by the water.

Most delicate and quite bulky cargo is transported via water. Such cargo includes goods such as petroleum and other organic substances. Such cargo vessels are regularly cleaned to avoid contamination of the cargo they transport. It is the duty of owners of such vessels to ensure that there cleaning services are done with the help of the best cleaning technologies available which prevent contamination of either surface water (for most cleaning services are carried out in seaports) it contamination of groundwater (for the cleaning water may infiltrate to deep levels in the ground) (Leonard, 2011).

Water bodies are used for different purposes. This is a fact that shipping companies should always consider before carrying out their activities in water. One is that they should ensure their water vessels do not cause pollute more so in terms of sound pollution produced by vessels engine. During construction of water vessels, the engines should be designed in such a way that they do not produce any noises during their operation. Noises do scare away the aquatic animals more so the fish from their breeding grounds. Recreational activities can no longer be done peacefully in noisy environments. Various water activities are distracted by noisy water vessels.

The use of surface water other than the general use should be granted permission by the water authorities. Such uses include the withdrawal of surface water for confiscation purposes. The use of such water in order to benefit from the water ability of generating electricity for personal use. Withdrawal of such waters for business activities such as breeding fish for sale, rearing various poultry breeds which rely on water for their survival among other reasons.

Use of groundwater should for other purposes other than the general uses should be granted permission by the water authorities. Such uses include the withdrawal of ground water to be accumulated for purpose of reducing its levels since during heavy rains it might be the main factor leading to flooding or any other use that boosts the economy (Leonard, 2011).

Not all times should permission for use of either surface waters or groundwater be granted. The withdrawal of water for Hydrological research activities which last for only a couple of days should not be granted any permission by the water authorities. Water accidents do take place all over the world. Sometimes there might be need of reducing the volume of water bodies in order to save victims. The reduction of water volume for rescue missions should not seek any permission for in most cases after withdrawal, the waters tend to be returned to their original sources. Permission should not be sought for training purposes of the marine personals.

Before the issuance of any water use permission, certain rules must be adhered to. The document presented should clearly state why the permission needs to be granted, the conditions for the water use or withdrawal among other requirements. If the permission being sought is for discharge of treated sewage water into a certain water bodies, it should not exceed a period of a certain number of years. This is because not all treated water goes through the required treatment procedures. Some water activities may take longer periods than expected, hence it allowed for application of extension of water use permission to be submitted to the water authorities (Leonard, 2011).

Most individuals apply for the water use permission in water bodies which are ever in demand by the general public. It is their responsibility to see to it that their personal use does not in any way hinder the normal water operations by the general public for such issues might lead to conflicts between the user and the public.

In situations where the authorized person is not within reach, he or she is entitled to allow a third party individual to run the water activities as stated in the permission document.

The carrying out of certain activities might affect out water sources in one way or the other. This means that before such activities are conducted, permission should be granted. However, before the issuance of such a permission, thorough research must be conducted to examine the damage which might be caused by such activities and the mitigation strategies which must be put in place to avoid such damage. Such activities include the planting of trees in wetlands which is a practice which might destroy the floodplains, however, it might promote the drainage system of such areas. Extraction of sand from water bodies does need application of extraction permission for if large amounts of sand are extracted, small river streams might dry up more so in drought periods. The disposal of previously withdrawn water from our water bodies needs permission for such waters must be tested to ensure that they will not end up polluting river streams (Carter, 2006).

All the permission document should be well recorded if possible electronically for future use. Any illegal activities such as issuance of bribe money should be discouraged completely for failure to adhere to anti-corruption rules, the individuals involved in such activities will have to face the severe consequences which have been put in place.

The CWA does state that it is a must for water sources to examined regularly to ensure that they are always in good condition. The examination conducted major on the quantity of water. They compare the current water volume in either surface water or groundwater water with previous results of the past. This ensures that the any changes in the earths hydrological levels are monitored. The quality of such waters should also be examined. Quality water test help the government in controlling the rates of pollution which reduces health complications among human beings and the other living organisms which depend of such waters for their survival.

Rivers and other small water bodies are known to originate at least from a certain location. The CWA states that such locations should be protected in the most appropriate way to avoid disappearance of river streams due to lack of enough water or contamination of the river streams at their point of origin. Activities such as cutting down of trees of in such areas is strictly prohibited under this Act. This is because help in increasing the precipitation levels hence enough waters is always available in the ground to feed such river streams. The extraction of raw materials and leaving the pits open is prohibited. This is because open pits tend to expose the groundwater to factors which might reduce its volume. The deposition of radioactive elements in water sources, be it the surface waters or the underground water is strictly prohibited. The effects experienced due to improper deposition of radioactive elements are quite diverse and irreversible (Leonard, 2011).

Some of the activities conducted by landowners might lead to contamination of groundwater. As we all know, most nations prefer groundwater to be used for drinking purposes only. This means that the contamination of groundwater will affect mostly human beings and cause various health issues. If the landowner caused the pollution willingly, he or she is required to compensate for the losses experienced. He could be required to cater for the treatment of groundwater. This means that the groundwater must be brought to the surface, treated, and then returned back to the ground through deep infiltration or through any other technique available.

Groundwater might increase its levels due to the ever changing climatic conditions in our surroundings. The rising levels might be noticed maybe during some minor activities such as farming activities among others. If any individual notices such, the CWA states that he or she is responsible for reporting the rising water levels to the appropriate authorities, more so the water authorities. Water authorities should see to it that immediate thorough research is conducted in such areas. This research should majorly focus on the cause for the rise of groundwater levels, the region which has been affected, the safety measures which must be put in place to avoid any damages to human life and their livelihood among other things. After the research, appropriate plans should be implemented immediately for the safety of everyone in such areas (Leonard, 2011).

Under the CWA, a section has been set aside to deal with issues to do with sensitive water areas. There are several sensitive areas which need to be handled with lots of care to avoid any negative effects. One of the sensitive areas is a water body whose quality has deteriorated completely to an extend that it will require huge amounts of capital to be said its clean. Even after thorough treatment procedures, such waters can never be considered to be safe for human consumption. This act states that the government should identify all sensitive areas and restrict them to avoid any more damages.

Vulnerable zones need to be protected too as stated in the CWA. Vulnerable zones consist of areas whose surface water or groundwater is designated only a particular, maybe for generation of hydroelectric powers, for drinking, for agricultural purposes among others. Such zones need to be identified and constantly monitored to ensure they remain in their original condition as before (Mittal, 2010).

Aquatic life needs to be protected. It is the duty of us as human beings to protect aquatic life for our biodiversity to be balanced. The CWA states that water bodies should be maintained in perfect conditions to preserve aquatic life. The fishing of indigenous fish species is completely prohibited for they contribute in the safety of our waters. Any release of new fish species in a certain water body, permission should be sought from the water authorities to ensure the fish species being released will be able to survive in such waters.

The amount of groundwater should be monitored from time to time, more so the amounts which are being withdraw constantly for human use in the earth's surface. Groundwater levels should not go below a certain level. The ground still needs water. Various underground living species depend on this water for their survival. If the levels are below the required level, the use of groundwater should be brought to an end (Leonard, 2011).

Waste water is a rising concern which needs to be addressed immediately. With the increase in population, flats are filling our urban centers. The waste from such locations has to be disposed off. The wastewater should be treated before release to our surrounding water bodies. The CWA states that human residences need to be equipped with wastewater treatment plans to avoid contamination of rivers.

The most dangerous type of waste is the hazardous waste. Such kind of waste is extremely harmful for it they lead to negative effects which might take ages for the public to recover from them. Under the CWA, it is strictly prohibited to release harmful waste to the surroundings. This is because they have the ability to decompose and sip into the groundwater contaminating them. Such waste should be gotten rid of completely either by incineration or any other way. They include poisonous chemicals, non-biodegradable waste, radioactive elements among others (Leonard, 2011).

When it comes to water rights, millions of rights have been implemented in order to help in this mighty fight of protection water in all ways, for as it has been said earlier, over, and over again, water is life.

4.9.3. Waters Protected Under the Clean Water Act

The CWA was implemented to protect all waters excluding none. From lakes, to rivers, to oceans and seas, to ponds, to floodplains, to wetlands, to artificial borings among others. Basically, CWA protects all forms of surface waters to ground waters.

Water Pollution Control Strategy: Since pollution activities are the ones endangering our water sources, various strategies have been out in place to help in reduction of pollution rates.

One of the strategies is the point source protection. This states that waste materials are handled carefully and disposed off in the most appropriate way at their point of source. Manufacturing industries are being equipped with appropriate tools to help in proper disposal of the waste generated during their manufacturing process. Such treatment activities ensure that water is not contaminated in any way.

Technologies are being innovated with each passing day to help in the journey of mitigating the effects caused by water pollution. Such technologies are more effective than human in carrying out their evaluation of controlling water pollution as compared to human beings. They produce trustworthy results and ensures records are maintained up to date automatically.

The normal uses of water from a certain water source should be well documented. This is to ensure that the government is aware that the waters of a certain water body are used for the following purposes so that, in situations where the waters are used for other uses other than the ones documented, legal actions are put into consideration.

Not all forms of pollution occur at the point sources. Water pollution taking place away from the water source is referred to as nonpoint source pollution. An example of nonpoint source is situations whereby fertilizers from agricultural fields are carried away by surface runoff into river bodies hence polluting the waters. Leakage of pipelines transporting organic substances such as oil might contaminate the underground waters if such waters sip into the underground levels (Stephenson, 2005).

Sewage systems in our cities need to be treated regularly before disposing off the dirty water in them into rivers. This preserves aquatic life and ensures that the waters are safe for human consumption.

Funding for Water Pollution Control Strategies: Appropriate techniques have been in place to help in the collection of the enough funds to facilitate the pollution control strategies. Clean Water movements have been formed to help in seeking sponsors are willing to help in reducing the pollution. Clean water tax is being levied on working individuals to ensure that they too play a role in keeping their waters clean. Regular fundraising events are being organized frequently to help in topping up money in the water pollution control accounts.

4.9.4. Challenges Facing the Clean Water Act

Despite all of its effort to preserve life in the universe, the CWA is being faced by so many challenges.

One of the challenges is the farming along river banks or in floodplains. To farmers, such grounds are known to be the most fertile surroundings due to deposition of river sentiments rich in nutrients. Such farming activities lead to the deposition of harmful waste products such as fertilizers and pesticides some of which are quite toxic and might render water bodies useless to all forms of living organisms. Rules and regulations are already in place, but despite the intensity of failure to obey such rules, people are still ignorant.

Not all individuals that contribute to the pollution of our water bodies are illiterate or aware of the rules and regulations which have already being put in place. The presentation of any pollution case with such an individual will only be a waste of time. Such illiteracy issues are caused by pack of enough capital to facilitate awareness campaigns to the general public.

Mismanagement of water pollution control funds is a major threat facing many nations. Individuals who have been put in charge to deal with the mitigation of water pollution are turning out to be embezzlers of funds which is quite a disappointment.

Corruption is threatening the good strategies which have been out in place to help in the reduction of water pollution rates. This is because unqualified personals are being given the responsibility of supervising the mitigation of water pollution effects and instead of doing their job well, they end up delivering shaggy results that are of no use to the general public.

4.9.5. Benefits of the Clean Water Act

Just like a coin has two sides, so does the CWA. Despite the challenges it has come across so far, it has been able to defy all odds and be of really great benefit to general public and the universe at large.

It has helped in maintaining our waters is perfect conditions for consumption. This has reduced the congestion of our hospitals with individuals suffering from illnesses caused by consumption of contaminated water.

Aquatic life is preserved for their survival environment are well protected and maintained in perfect conditions. This ensures that our biodiversity is balanced. Fish which is used as human food is made available since more fish species will be able to thrive in such perfect conditions.

Recreational activities are conducted with the thought in mind that the waters are secure. Human are more entertained by the various water sports.

In conclusion, the CWA has not come into action recently, it is only that, for ages, it was part of other environmental laws until water pollution rates become a threat to the entire globe. This rates brought into attention the dangers ahead and pushed many governments to come up with laws that will majorly focus on the protection of waters. It is our responsibility to ensure that we adhered to all the laws which have been put in place to protect our waters. This is because, we all need water for survival since water is us and we are part of water. Amendments should be done frequently to increase our chances of overcoming the water pollution issue.

4.10 WATER MANAGEMENT

4.10.1 The Safe Drinking Water Act (SWDA)

Introduction

In the United States of America drinking water that is available to the public is governed in quality by the rules and regulations and acted by both federal and state governments. At a local level there are also sitting injunctions that may be created however the fundamental federal law is known as the SDWA. The SDWA cannot be Express without be turning out the United States Environmental Protection Agency (US EPA) that is authorized to design create and enforce regulations to reach the SDWA cold. Environmental Protection Agency (EPA) is an agency that is independent and executive and essentially part of the United States federal government which governs Environmental Protection. It was created due to the increasing public concern about human activity and the impact that it has in the environment in the late 50s and into the 60s. Overall the detrimental effects on the environment due to some indiscriminate practices eventually introduced bills over time that eventually passed through the house-senate colloquiums concrete and implementation of national environmental policies (Pontius, 2003)

Public water systems that are entities who provide water that is meant to be fit for human consumption through pipes and various constructed conveyances that serve an average of at least 25 people for a minimum period of 60 days a year and essentially describes not the ownership of the system but the people who have access to the water from it as in the public. The SDWA was first enacted in 1974 passing through congress to regulate the Nation's public drinking water supply and went through amendments in 1986 1996 and 2016. Programs that set standards and treatment requirements for public water supply, financing of infrastructure projects that supply drinking water and overall water system compliance is administered through this act and the bodies that enforce it which are both local and federal governments with Bodies Like the EPA. Most States have public water supply supervision programs that regulate public water systems who are authorities in implementing this act and generally the 1974 law established arrangements where the States were placed in charge in deciding who is delegated for primary implementation and enforcement (Pontius, 2003).

When looking at the legislative history it is democratic Senator Warren Magnusson from the state of Washington that first introduced the drinking water act in January of 1973 in which the Senate convened and passed a version of The Bill six months later in the same year. Legislation also has to go to the House of Representatives and in November 1974 they passed their own version of The Bill with the Senate again amending the house bill before passing it again. When disagreement was reached between the house and the Senate the bill was put into law on the 16th of December 1974 by President Gerald Ford who signed it. Upon signing it President Ford made several remarks including its purpose in the establishment and enforcement of national drinking water standards and EPA's responsibility in that regard (Trimble, 2011).

4.10.2. Background of the Act and Various Amendments

The first standard set by the SDWA were to monitor public water systems and some inorganic contaminants like metals and nitrates bacteria and other microbial contaminants and trace amounts of radioactive materials like uranium with amendments being subsequently followed. The 1986 amendments had 83 specific contaminants put into regulation and the amendments made in 1996 where in the restructuring of appropriation for programs that were under the SDWA and also placed the EPA in charge carrying out of cost benefit analysis over drinking water standards true to the end of 2003 fiscal year. Although will cover many of these in detail later in this chapter the Table 4.1 is an idea of the original enactment and the subsequent amendments.

Table 4.1: Safe Drinking Water Act and Amendments (codified generally as 42 U.S.C. 300f-300j)

Year	Act	Public Law Number
1974	Safe Drinking Water Act of 1974	P.L. 93-523
1977	Safe Drinking Water Act Amendments of 1977	P.L. 95-190
1979	Safe Drinking Water Act Amendments	P.L. 96-63

1980	Safe Drinking Water Act Amendments	P.L. 96-502
1986	Safe Drinking Water Act Amendments of 1986	P.L. 99-339
1988	Lead Contamination Control Act of 1988	P.L. 100-572
1996	Safe Drinking Water Act Amendments of 1996	P.L. 104-182
2002	Public Health Security and Bioterrorism Preparedness and Response Act of 2002	P.L. 107-188
2011	Reduction of Lead in Drinking Water Act	P.L. 111-380
2013	Community Fire Safety Act of 2013	P.L. 113-64
2015	Drinking Water Protection Act	P.L. 114-45
2015	Grassroots Rural and Small Community Water Systems Assistance Act	P.L. 114-98
2016	Water Infrastructure Improvements for the Nation Act	P.L. 114-32

Source: https://fas.org/sgp/crs/misc/RL31243.pdf.

As shown above the original act was to first give the EPA discretion and authority to regulate the contaminants that exist in drinking water with States playing the major role in implementation and enforcement. The first major amendments to the act were made in 1986 to increase the pace in which EPA regulate contaminants and also to make concrete the protection of groundwater. Surprisingly between 1974 and 1986 only one more contaminants was added to the regulation by the EPA behind the 22 standards that were developed by the public health service prior to that. The 1986 amendment required the EPA to set up regulations for 83 specific contaminants by 1989 and 25 more contaminants for every subsequent year for three years after that. This together with promulgation of requirements for filtration systems and disinfection levels and the reduction of the use of lead pipes and lead solder were also very important (Tiemann, 2011).

The 1986 amendment also established protective programs for public Wells especially with the Wellhead systems that they used. A Wellhead is a component at the surface of a well that provides the pressure for pumping of water oil or gas underneath the ground and specifications were put in place on how they should be managed. Furthermore, the acts put in place a grant program for both state and local authorities where aquifers were the primary source of water and developed groundwater protection programs through the amendment. An aquifer is essentially an underground layer of rock that is permeable and water bearing and the amendment set in place regulations in this respect. Finally, 1986 amendment was also ranked as the first major adjustments in the SDWA in that rules for monitoring underground injection Wells that inject hazardous waste below underground drinking water sources were issued which basically means the EPA's enforcement authority was really backed up.

The SDWA was amended by Congress again in 1988 through the lead contamination control act which was intended to reduce exposure to levels of lead in drinking water by recalling all lead-lined water coolers as a provision and reinforcing the EPA and as an authority by instructing them to issue a guidance document and testing protocols at the state level. When the 1986 amendments proved themselves to be unworkable for the EPA public water systems and the States generally Congress had to step in and design new regulations within the SDWA in 1996. The 1996 amendments went along the lines of setting new appropriations for federal drinking water mandates increasing the regulatory flexibility for the amendments to be workable as well as directing the resources towards reducing the greatest health risk. This risk-based approach made sense and inspired Congress to rectify the requirement that the EPA had to place 25 contaminants after every three years into regulation (Pontius, 2003).

In discussing regulatory flexibility this was mostly in relation to the standard-setting process the EPA had to carry out to conduct health risk and cost analyses. In order to help water systems finance infrastructure projects that fall in line with the SDWA regulations the EPA was also authorized to set up a Drinking Water State Revolving Loan Fund. These and other programs where established to improve compliance levels focusing on the acts requirements on pollution prevention communication on consumer information and other assessment prerequisites. The 1996 amendments where to run through the end of the 2003 fiscal year.

The public health security and bioterrorism preparedness and response act of 2002 hard water security provisions that required Amendment of the SDWA. This sat requirements for community water systems that served more than 3300 people and stipulated vulnerability assessment to be conducted together with the design of emergency response plans. Criminal and civil penalties were increased in legislature for those who tampered with the water supplies and so the EPA was required to carry out research on preventing and responding to terrorist attacks (Trimble, 2011).

The reduction of lead in drinking water act was signed in 2011 and made stricter the SDWA definition of lead-free water. The act now banned the use and sale of lead pipes and any plumbing fitting of fixtures that were not deemed to be lead-free. An amendment was made that specifically exempted fire hydrants from the lead plumbing restrictions that the act covered which would have been impractical if it did. 2015 saw the passing of the drinking water protection act that brought about new regulation around algae toxin in public drinking water supplies which was implemented by EPA that was instructed to design a strategic plan to manage the risks associated. Later towards the end of the same here the grassroots rural and small community water systems assistance act was also enacted that regulated small technical systems assistance programs and also extended the funding through to the end of the 2020 fiscal year. Congress made several amendments in the SDWA through the waste water act in 2016. Through this act Grant programs were established most specially to service public water systems in small or disadvantaged communities for them to be able to meet the SDWA standards and regulations. It also established a program that focused on lead testing in drinking water in schools and overall other lead reduction projects.

4.10.3. Public Water Systems as Described by the EPA through the Safe Drinking Water Act

Federal regulations that apply to drinking water pertain to more than 152,000 public and private water systems the regularly serve a minimum of 25 people with water for human consumption. One can imagine how diverse is water systems are and the very in both size and type. 51,350 of these public water systems fall under the category of community water systems that provides safe drinking water for the same residence all year round. These community water systems serve around 300 million people and all federal rule and regulations apply to them. Majority of these community water systems are fairly small serving 3300 individuals or less. These accounts to up to 82% of them however they serve only 9% of the total population as described above. 92% of community water systems serve populations of 10,000 people or less with 8% that serve population greater than 10,000 providing water to 82% of the population which accounts for more than 246 million individuals. Another interesting statistic is that majority of these community water systems 71% rely on groundwater and 29% of them on surface water (Carter, 2006).

Apart from community water systems, there are also non-transient non-community water systems that account for 18,000 178 public water systems and these include schools or factories that have their own individual water supplies serving the same people not all year round but for more than six months. Not all federal regulations on the SDWA apply to them but many do. They are small with 99% of them providing water to 3300 people or less.

There also exist transient non-community water systems under public water systems and their 83,200 of them. These account for such areas as campgrounds or gas stations which supply water to the momentary or temporary population. The regulations from the SDWA that apply to them are mostly those concerning contaminants that may pose health risks. These statistics are important when looking at the SDWA in the financial managerial and technological perspective and the challenges that may come up in meeting federal regulations for drinking water. And below is a table that provides these statistics on community water systems non-transient non-community water systems and transient non-community water systems. The reason why they are broken down like this in type is as a function of the EPA and they are also the source of the stated statistics (Carter, 2006) (Figure 4.1).

		Water System Population Size Category					
		Very Small 500 or less	Small 501-3,300	Medium 3,301-10,000	Large 10,001-100,000	Very Large >100,000	Totals
CWS	# Systems	28,346	13,737	4,936	3,802	419	51,356
	Pop. Served	4,763,672	19,661,787	28,737,564	108,770,014	137,283,104	299,216,141
	% of Systems	55%	27%	10%	7%	1%	100%
	% Of Pop.	2%	7%	10%	36%	46%	100%
NTNCWS	# Systems	15,461	2,566	132	18	1	18,178
	Pop. Served	2,164,594	2,674,694	705,320	441,827	203,000	6,189,435
	% of Systems	85%	14%	1%	0%	0%	100%
	% Of Pop.	35%	43%	11%	7%	3%	100%
TNCWS	# Systems	80,347	2,726	92	13	1	83,179
	Pop. Served	7,171,054	2,630,931	514,925	334,715	2,000,000	12,651,625
	% of Systems	97%	3%	0%	0%	0%	100%
	% Of Pop.	57%	21%	4%	3%	16%	100%
	Total # of Systems	124,270	19,029	5,160	3,833	421	152,713

Figure 4.1: Public water system statistics in terms of water systems regulated under the SDWA.

Source: http://water.epa.gov/scitech/datait/databases/drink/sdwisfed/upload/epa816r13003.pdf.

4.10.4. National Drinking Water Regulations

To the SDWA the EPA is required to disseminate national primary drinking water regulations especially for contaminants that could potentially pose health risks. These federal standards are applicable to all water-systems whether privately or publicly owned that regularly service 25 people at the least and the EPA regulates these contaminants and select they based on:

- Potential adverse health effects that they pose;
- High probability that the contaminant will be found in the public water systems at a level and frequency that could negatively impact public health; and
- Likelihood that selecting and regulating such contaminants reduce health risk for regular users of public water systems.

The EPA is required to publish a list of unregulated contaminants that is expected to appear in public water systems that will need to be regulated every 5 years. This list is known as a contaminant candidate list. The EPA is further responsible for administering a monitoring program under the SDWA that would facilitate the collection of data in places where the contaminants are suspected to be present and publish an unregulated contaminant monitoring rule that lists no more than 30 unregulated contaminants that are to be monitored also every 5 years. There is also a requirement under the SDWA for the EPA to review at least five on the contaminants included in the contaminant candidate list whether they should be regulated or not (Carter, 2006).

Non-enforceable maximum contaminant level goal must also be set by the EPA for every contaminant that they feel requires regulation so that this contaminant maybe at a level that no adverse health effects will be experienced and also set maximum contaminant level which is enforceable close to the non-enforceable levels but with an adequate margin of safety that is deemed feasible. Once the EPA's administrator determines that a contaminant must be regulated they proposed a rule within 24 months and circulate in national primary drinking water regulation within 18 months after the proposal.

If there are potential health risks even though the feasible levels of a contaminant the EPA can set another standard outside of this on concentration levels of other contaminants. In this case there must be a standard or treatment technique used to minimize overall potential health risks. Also, when suggesting a new regulation, the EPA has to circulate a determination on whether or not the benefits of this new standard will justify its implied costs. If the benefits do not justify the costs the EPA wood communicates a standard that maximize as health risk reduction benefits while reducing costs (Glavan, 2018).

Historically legislations that are set needs the EPA to base them off of technologies that r cost-effective for large communities however the 1996 amendments of the SDWA requires the e p a to list any technology or various methods that compliance levels will be met especially in relation to the minimum contaminant level even for small public water systems that serve 10000 people or fewer. This includes a small sister of variance technology that are in some way shape or form protective of public health if the compliance technologies are not identified by the EPA. The EPA is required to review, modify extend and strengthen where appropriate each of the drinking water regulations after every 6 years and new regulations are usually put into effect three years after the initial promulgation it is also important to mention that for some contaminants the EPA is unable to place a minimum contaminant level and instead establishes at treatment technique instead these federal drinking water standards divided into six groups:

- Microorganisms;
- Disinfectants;
- Disinfection byproducts;
- Inorganic chemicals;
- Organic chemicals; and
- Radionuclide's.

EPA established national primary drinking water regulations that fall under the code of regulations in the United States and this is why the authority of the SDWA and it is very definitive in its description of what a contaminant is. A contaminant is described as any physical chemical biological or radiological substance or matter in water which may appear very broad but it is anything other than water molecules. In any case drinking water reasonably should be expected to have at least small amounts of some sort of contaminant and what's the SDWA is important for is in putting the EPA in charge of identifying contaminants that may be harmful if consumed at certain levels because many others are harmless. In fact, if anything the EPA has made it clear that the presence of contaminants does not necessarily mean that there is a potential health risk being posed by consuming the water (Glavan, 2018).

Microorganisms: EPA through the authority mandated by the SDWA has standards all sorts of bacteria and viruses and requires two microorganism related tests to be done to indicate the quality levels of drinking water which are plates count and turbidity. Play counts or a type of bacteriological water analysis methods where we observe bacteria growing a colony on a medium and eventually the Colony will become visible to the naked eye and then they are counted. These testing methods to be effective have to use a dilution of the original sample so that the target bacterium is grown. For margins of error and all scientific purpose is usually more than one dilution is cultured and this approach is widely used for measuring the effectiveness of water treatment.

Turbidity is the other measure and it refers to the cloudiness or haziness of a fluid as a result of contaminants four particles which can be visible to the naked eye the same way smoke is in the air and is a key test of water quality. The most popular scientific instrument used to measure turbidity or haziness is a formalin turbidity unit which even has its international organization of standardization measuring unit known as nephelometric units. There also exists the ISO 7027 that is a standard for water quality measurement, an Arduino micro controller that can be used with LEDs to measure the same (Pontius, 2003).

The initial surface water treatment rule in 1989 and the long-term to enhanced water treatment through of 2006 are the two major legislations put forward by the agency towards microorganisms in water. Dementia 2006 promulgation was important because it required public water systems to use a treatment technique to control cryptosporidium and other pathogens. Below is a list describing the various microorganism contaminants and the maximum contamination levels according to the EPA with most using treatment techniques to get rid of them. Microorganisms that should be regulated include:

- Cryptosporidium;
- Giardia lamblia;
- Heterotrophic plate count;
- Legionella;
- Coliforms;
- Turbidity; and
- Viruses.

Disinfectant and Disinfection Byproducts: The below list illustrate the disinfectants and disinfection byproducts that are stipulated by the EPA that have shown to cause cancer and reproductive effects in lab animals and suggest bladder cancer and reproductive effects in human beings as well with their respective maximum contamination level goals and MCLs shown. Under the act of SDWA they are 81 disinfectants and disinfection byproducts rules that were established as federal regulations in 1998 and Stage 2 disinfectant and disinfection byproducts rule that were placed into federal regulations on January 2006. The disinfection byproducts rules apply to all sizes of community water systems and non-trans and community water systems that use some sort of disinfectant other than ultraviolet light to deliver water that is clean. The rules also apply two transient non-community water system that use chlorine.

- Chlorine dioxide (as ClO_2);
- Chlorine (as Cl_2);
- Chloramines (as Cl_2);
- Bromate;
- Chlorite;
- Haloacetic acids (HAA5); and
- Trihalomethanes (TTHMs).

Inorganic Materials and Chemicals: The 1986 amendments to the SDWA extended the EPA's authority to fix standards that limit the concentration of lead as stated above and the 1991 regulation added copper to the list that if present in public water systems would be potentially harmful to human health and therefore have specific standards issued for them. Chemical contaminants in the rules around them communicated in phases do the chemical contaminants rules Phase II though V. The chemical container rules were stipulated to regulate over 65 contaminants that fell on the three groups.

- Inorganic contaminants;
- Volatile organic contaminants; and
- Synthetic organic contaminants.

What are organic contaminants and synthetic organic contaminants will be shown in the immediate next section however they were all treated in a similar fashion with reference materials like technical fact sheets and drinking water standard for arsenic fact sheet. They also dictate the long-term risks like cancer organ damage circulatory system disorders nervous system disorders and reproductive system disorders that are potential health risks if these chemicals are present in drinking water systems (Pontius, 2003).

Organic Chemicals: By definition organic chemicals is a variety of substances that contain carbon in a covalent bond in many of its derivatives. They are used in many different Industries in a wide variety of applications. Some are used to make feedstock while others are used as ingredients in polymer plastic production and the solvents and cleaning agents. Organic compounds are also found in living things in the form of carbohydrates lipids proteins and nucleic acids however there are some that are harmful for consumption and therefore cannot be present in public water systems and the SDWA.

4.10.5. Variances Exceptions and Oversight in the Safe Drinking Water Act

One of the other requirements under the SDWA is taking into account smaller public water systems that may have difficulty in complying with every single regulation and for this congress allowed variances and exceptions in the SDWA as a provision. These variances are granted at the state level as they are authorized in Section 1415; however, it must be in spite of application of the best technology. If EPA cannot identify some technologies that cannot fit into or are too expensive or smaller water systems is in the EPA allows exemptions. It is important to note that under this subsection of the SDWA variances are not available for microbial contaminants because regardless of how large or small the water system is they pose a threat on human health and they are treatment techniques to rid the water of these contaminants which are all affordable compliance TECHNOLOGIES (Bos, 2016).

Exemptions are a little bit different as under section 1416 states are authorized within the jurisdiction to make temporary exemptions for public water systems treatment techniques or standards if they cannot be complied with four reasonable and compelling reasons including costs. This is intended to allow more time for a given public water system to meet regulation requirements as long as it will not result in any unreasonable health risk period and exemption for a water system can be issued with a deadline of not more than 3 years after the compliance deadline and systems lots of 3300 people or less can receive another three additionally years together with an additional 2 years' extension with the maximum extension deviation being 9 years (Bos, 2016).

Oversight of public water systems is under section 1413 of the SDWA that also is not only States but Native American tribes in the reservations to assume primary oversight and enforcement responsibility that is legally quoted as primacy however they must meet statutory requirements set by the EPA. For privacy to be assumed for state they must first adopt regulations at least as strict as those on the national level and secondly they must develop appropriate procedures to enforce regulations like conducting monitoring and inspections. They must also have an authority or administrative body that is able to dish out penalties and conduct Inventories of water systems. Furthermore, States are also meant to design a strategic plan for the provision of safe drinking water under emergency situations. Through 1997 and 2300 school years Congress authorized 100 million dollars annually for public water supply supervision programs in every state and the EPA is responsible for making grants to the States. The way the EPA is to do this allotment is following the basis of population geographical area number of Public water systems and various other relevant and pertinent factors. This together with the Drinking Water State Revolving Loan Fund both serve as adequate financing to ensure that public water supply supervision programs are well administered.

4.10.6. Enforcement of the Safe Drinking Water Act

The SDWA is a legislation that was designed in a way that the public water systems our first responsible for monitoring their own water supplies and ensuring that they meet compliance levels by reporting all the monitoring results to the states. The States where these drinking water supplies are then review the monitoring data submitted but also carry out their own monitoring to make sure that drinking water regulations are met. The EPA will review the violation data submissions (Trimble, 2011).

This three-pronged approach is also reviewed in section 1414 of the SDWA whereby if the EPA is that a public water system does not comply with regulations then they report back to the States and the public water system providing assistance to improving the water supply to regulation levels. The state is then responsible for commencing enforcement and if they are unable to within 30 days after the EPA's notification then the EPA is required to carry out an administrative order or proceed with civil action. States that do not have privacy which only include two of the 57 states and territories in the United States wheel need an elected official at the local level to be notified before any enforcement action is taken against the public water system.

SDWA amendments made in 1996 gave enforcement authorities more strength. The EPA's enforcement team was also streamlined and so where the processes that they had in issuing federal administrative orders. Administrative penalty amounts were also increased making the regulations more stringent. Many sections in the act were made clear and enforcement action was limited towards public water systems that were making plans to consolidate with another system. This among other reasons is why we keep repeating the 1996 as one of the most important amendments in the SDWA (Trimble, 2011).

The consumers of water from these public water systems are the primary reason why the SDWA exists and that is to protect them. This means that when public water systems do not meet regulation standards then they are to inform or notify their customers of such violations. Section 1414 of the SDWA states that systems must notify customers within a 24-hour period of any violations that could have potential adverse health effects that are serious to consumers. The Water Infrastructure Improvements for The Nation Act of 2016 also includes public notification requirements in section 2106 especially in relation to lead levels under the EPAs lead and copper rule. This is important because notification requirements in the past only applied to violations of standards and not to exceedances of Lead action levels. This law was also restructured such that exceedances that are potentially hazardous to human health is supposed to be reported by the public water system to the public the state and the EPA within 24 hours. Because the state the EPA and the public water system with all have the violation or exceedance information at the same time, the public water system does not provide the required notice the EPA must again within 24 hours promulgate such information to the public (Carter, 2006).

Areas that have been affected by lead in their water systems also need education and technical assistance provided by the EPA. The Water Infrastructure Improvements for the Nation Act had its impact on the SDWA in the EPA is required to design community outreach programs, making sure that if a household indicates exceedances action by the state should be carried. There also exists an annual consumer confidence report that community water systems are to provide to all customers which the EPA uses to prepare (as well as other pieces of information) their own annual national compliance reports.

Civil action through civil suit can be carried against any person or agency that violates sections of the act and even against the EPA administrator in the event that they fail to perform any duty that is not discretionary. This gives the customer power and legal responsibilities as well in this respect because everyone is to some extent responsible for the water supply including the consumers themselves (Carter, 2006).

If by any chance a contaminant is likely to enter the public water supply and affect the drinking water in a way that has potential adverse health effects and the state or local authorities has not gone ahead to make appropriate moves, then the Administrator of the EPA is authorized under section 1431 with emergency powers to issue orders and even carry out civil action.

4.10.7. Lead in Drinking Water under the Safe Drinking Water Act

Lead enters drinking water mostly through plumbing material and water container transport systems that will bring the water to the household. It mainly leaks through lead corrode especially the water is acidic all the waters mineral content is low which ends up corroding the pipes fixtures or lead containers. Lead service lines are the biggest features of lead greenhouse hoops that can contaminate water include pipes and faucets as well as fixtures. Due to the SDWA modern household or cities in general are less likely to have lead pipes the older cities or homes built before 1986. Brass or chrome plated brass faucets or general plumbing that are built with lead solder which is essentially an alloy that melts at low temperatures is a feature that the SDWA up until today he's still trying to reduce and phase out (Tiemann, 2011).

According to the SDWA section 1417 lead-free plumbing and other associated material that comes into contact with water is approved and anything weighed over 0.25% across the wetted surfaces and 0.2% in lead solder is unacceptable and against regulation. The act has a process in which weighted surfaces are calculated and averaged out.

Lead gets into drinking water through a process known as corrosion that is when the metal wears away or dissolves due to chemical reactions. The rate at which lead grows into water is dependent on several factors including;

- The pH level of water whether it is leaning towards the acid or alkaline sides;
- The kind of minerals the water contains and the concentrations in which they are present;
- The total amount of lead that the water comes into contact with;
- The temperature of the water when it is in contact with lead surfaces;
- The wear and tear along the pipes;
- The period of time that the water stays in the pipe; and
- The coating or protective skills that exist inside the plumbing.

The lead and copper rule which we have previously mentioned is how the EPA by the authority given to them with the SDWA is how this issue of corrosion of lead and copper is dealt with. This does not mean that lead and copper cannot be used for plumbing fixtures pipes and such but they are required to be treated for corrosion control before usage. Because of the way the EPA designed their maximum contaminant level goals they are mandated to try and figure out if they are actually safe levels in which these contaminants can be present in the drinking water without potentially causing adverse health effects. Lead levels are set at zero by the EPA because it is a toxic metal which even at low exposure levels can be dangerous to human beings and they are bio accumulative in nature. The most vulnerable would be the Young and the unborn. A dose of lead that can be of little to no effect to an adult can actually have significant and even life-threatening effects on a child and they damage the central nervous system causing learning disabilities stunted growth and impaired blood cell functions (Glavan, 2018).

Reductions in lead exposure throughout the United States have been monitored through mass testing of blood lead levels over time even in epidemiology studies conducted by the center for disease control and shows that since the 1960s there has been increasing lower levels of lead like shown on the table below. Exposure however must be noted that it doesn't come from drinking water alone but exists also in paint dust soil even in the air. Section 1417 that we have already discussed was added in 1986 (Figure 4.2).

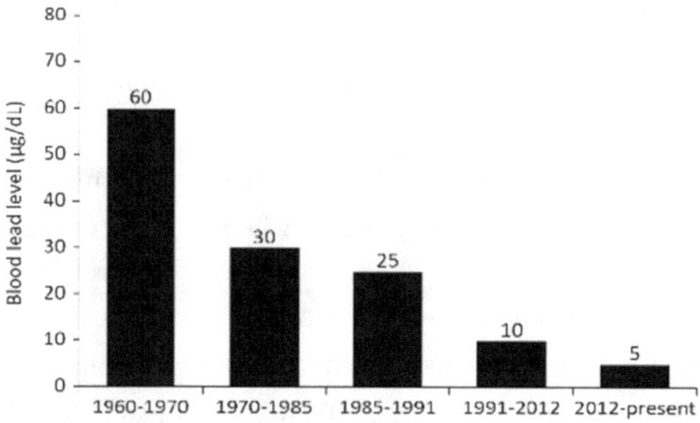

Figure 4.2: Blood lead levels monitored over time.

Source: https://www.atsdr.cdc.gov/csem/csem.asp?csem=34&po=8.

In 1996 Congress added another subsection to section 1417(e) instructing the EPA to implement regulations based on health standards that would further limit lead levels that could leach or corrode as a result of new plumbing fixtures and this was developed into an industry standard. (Tiemann, 2017) Available at: https://fas.org/sgp/crs/misc/RL31243.pdf In 2011 congress pass the reduction of lead in drinking water act that further amended this section in the SDWA and modified the definition of lead free prohibiting the sale or use of any lead-based material for plumbing fixtures and fittings. This is when the weighted average was dropped from 8% to the above stated 0.25% but included some exceptions like irrigation systems industrial processing equipment. Three years later in 2014 the provisions of the act became effective an existing regulation well then modified and revised by the EPA to confirm to the reduction of lead drinking water act (Glavan, 2018).

Under the water infrastructure improvement for the Nation's act in 2016 the EPA is authorized to provide information on lead in drinking water widely available to the public and this act also replaced the SDWA subsection 1464 on testing lead levels in drinking water in both schools and child-care programs where Congress appropriated 20 million dollars per year between 2017 to 2021 fiscal years.

4.10.8. Compliance Capacity and Development Programs and Operator Certification Programs

The operator certification program and the capacity development program are two state-administered programs that are designed to improving public water systems and their compliance to the SDWA regulations that were included through the 1996 amendments. Schools and workplaces with their own water systems that are non-transient non-community water systems are also to be trained through state adopted programs on regulations and how to abide by them. If a state does not adopt and implement the operator certification program the EPA can actually withhold the Drinking Water State Revolving Loan Fund by 20% for this non-compliance. The state is further required to have capacity development programs that have a legal authority to ensure new systems have oil capacities in terms of finance managerial and technical dimensions so that they can meet the SDWA regulations together with a strategy to aid water systems that already exist but experiencing trouble in reaching compliance levels. Lack of a compliance capacity and development program will mean that the EPA is authorized to withhold the state revolving fund grants from the state. So far the agency has not had to take such measures in terms of withholding funds for either of these programs (Pontius, 2003).

Other programs exist to complement the above two state programs and they include compliance assistance programs authorized by the EPA and the States especially for small public water systems that serve 25 to 10,000 customers. This is because these community water systems that are small lack economies of scale and various capacities to meet the act's requirements. This was included in section 1442 of the 1996 amendment and between 1997 to 2003 fiscal years Congress appropriated 15 million per year for these organizations that were non for profit. The grassroots rural and small community water systems assistance act was enacted in 2015 after being passed in 2012 which extended this appropriations authorization through to the end of the 2020 fiscal year. They include anything and everything from assistance in implementing the regulations to protecting the source water and providing monitoring plans. Another source of funding came through the Drinking Water State Revolving Loan Fund for technical assistance particularly to assist small systems.

For small and disadvantaged communities Grant assistance also comes to The Water Infrastructure Improvements for the Nation Act for the Nation's act that instructs the EPA to design Grant programs specifically to assist disadvantaged communities that are unable to finance projects that would adhere to the SDWA regulations. In fact, the EPA is supposed to prioritize underserved Communities when it comes to funding by investing in households water quality tests and any sort of assistance that would benefit these small and disadvantaged Communities on a per household basis.

4.10.9. Protection of Underground Sources of Drinking Water and Underground Injection Control Program

Almost all public water systems in the USA depend on groundwater especially aquifers that we earlier explained are water-bearing permeable rock which is extracted for water through such things like wells and boreholes. The SDWA prohibits any underground injection that would damage underground water sources. There are state established programs that are entitled to underground injection controls programs and protection of underground sources of drinking water true mandated regulations of the EPA promulgated in 1977 (Bos, 2016).

Legally what is required to happen under section 1422 of the safety of CWA is that states are supposed to submit their own designed plans to the EPA for the implementation of underground injection control programs where they will be approved and from there they are to assume primary enforcement and responsibility of these programs. If the state submits a plan that is not approved, then the EPA must step in and design and implement a program on their behalf period because of the energy policy act of 2005 the SDWA was amended so that underground injection includes injections of fluids used in fracturing operations like brine from oil and gas Productions.

Underground fluid injections could have extremely dangerous consequences on groundwater drinking sources and by extension adverse effects on human health. It is also very difficult for injected fluids to be traced once it enters the ground and begin to contaminate the underground sources of drinking water. Remediating polluted underground sources of drinking water is even more difficult and therefore a preventative approach was implied by congress with the SDWA and States with an underground injection control program are authorized with primary enforcement for Oil and Gas injections even though we don't meet EPA's if and only if the program effectively prevents underground injection from endangering drinking water sources (Bos, 2016).

This implementation and enforcement primacy has been issued for all kinds of boreholes and Wells to 35 States by the EPA. Implementation and enforcement responsibility is shared with the remaining nine States and also with two Native American tribes for underground injection control programs.

Full implementation of the EPA inverter rated six classes of underground injection control Wells based on the likeness of the fluids being injected construction depth of injection design and general operating designs and techniques. The EPA also issued directives for the customary performance criteria in each of these six classes. Latest issuance of ordinances in respect to this subsection was for class 6 wells which stipulated requirements for underground injection of carbon dioxide. Injection of carbon dioxide on the ground is a way to prevent emissions of greenhouse gases (GHGs) from coal plants that produce electricity. The other underground injection control systems of wells are listed below in accordance to their classes (Bos, 2016).

- Class 1 industrial hazardous and non-hazardous waste including wastewater from municipal sources and disposal Wells.
- Class 2 oil and gas injection wells with the exception of hydraulic fracturing.
- Class 3 solution mining world that are in situ leaching or in-situ recover.
- Class 4 injection wells dealing with radioactive wastes which is no longer permitted.
- Class 5 wells that deal with non-hazardous fluid injection into or above underground sources of drinking water.
- Class 6 injection wells for carbon dioxide for the purposes of geological sequestering.

There are four other programs mandated by the state EPA through the SDWA to categorically protect underground sources of drinking water and they are listed below:

- The sole-source aquifer Protection Program that was included in the SDWA of 1974 that protects the land on top of aquifers that are the sources of drinking water in the area and countrywide there are 77 sole source aquifers designated by the EPA.
- The sole-source aquifer demonstration program that was included in the SDWA Amendment of 1986 that officiated procedures for demonstration programs 4-hour fire protection areas.
- The state Wellhead protection programs that was also included in the 1986 amendment that initiated elective state programs for the

Protection of Wellhead areas that supply public water systems. If the EPA approvals the state's Wellhead Protection Program for a particular state, they can award grants of up to 90% of the implementation costs for the program.

- The state's groundwater protection grants that ratify the EPA to award 50% to state that ensue programs that protect groundwater is within their states.

Congress authorized appropriations of 15 million per year through the 2003 fiscal year and the Drinking Water State Revolving Loan Fund given to which state can be used partly in the groundwater protection activities.

4.10.10. Source Water Assessment and Protection Programs as Described by the Safe Drinking Water Act

As much as underground sources of drinking water are the major source of drinking water in the United States other sources like rivers and Lakes and any other place we nay get drinking water through the natural water cycle are also assessed and protected under the SDWA and derivative programs. Source water refers to all of them including aquifers and as much as that we have already discussed them and congress's approach to the development of the SDWA meant using pollution prevention perspective for the source water protection. This protection approach of prevention of contamination of water sources also contains a multi barrier focus. This happens by providing methods to prevent contamination of the source water using appropriate water treatment systems using apposite distributed systems water testing and even education and training programs for water administrators and managers. It can be up to 40 times more expensive to remediate source water contamination then actually taking steps in the beginning to protect the water in the first place. On top of that the united nation has called for current trends of wasting and pollution to be halted because there's not in as little as two decades two out of three people on earth will suffer moderate to severe water shortages (Figure 4.3).

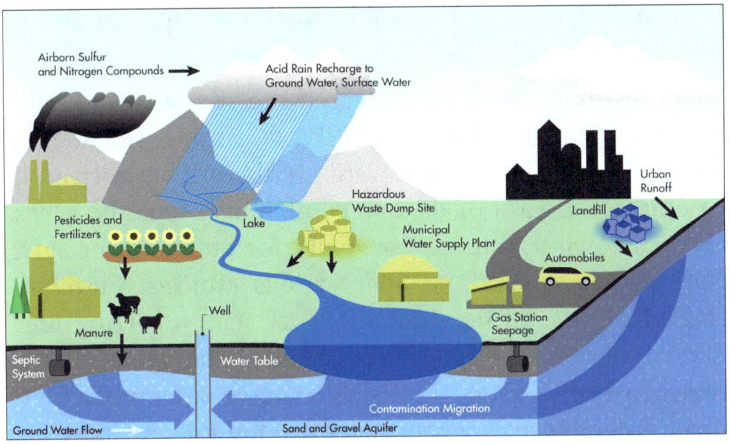

Figure 4.3: How human activities affect source water.

Source: https://www.safewater.org/fact-sheets-1/2017/1/23/source-water-protection.

States and local governments are meant to cooperate and the SDWA to develop source water protection programs and establish in working relationship to reduce the existence of contaminants in drinking water organize financial and technical efforts and develop a long-term source water protection strategy. Source water protection programs r appropriated 5 million dollars each year for grants and States can use as much as 10% maximum of their drinking water state revolving fund Grant for various activities that support source water protection and that includes the petition program. These programs help prevent both point source pollution and nonpoint source pollution.

4.10.11. State Revolving Funds under the Safe Drinking Water Act

The drinking water state revolving fund was authorized by congress in 1996 and this financing is meant to be directed towards making improvements across the boards to comply with regulations of the SDWA. The EPA is in charge of the state revolving fund and must make sure States capitalize on the usage. States are also to match 20% of this national Grant which is ordinarily allotted on the basis of surveys done by the EPA based on the needs each state manifest. Each state together with the capital city District of

Columbia must receive at least 1% of the total funds appropriated (Trimble, 2011).

The state revolving fund may be used broadly even to provide loans for various expenditures as long as the EPA determines that the action will significantly play a positive role towards furthering the act and its health protection objectives. 15% of the annual allotment must be accessible by each state for known assistance especially for public water systems that serve 10000 people or less and their various projects. Furthermore, the SDWA states that 30% of the drinking water state revolving fund can be used to service the economically disadvantaged communities and provide loan subsidies to them. Technical assistance source water protection capacity development programs and operator certification programs are all eligible to funding from the state revolving fund Grant as authorized by the state (Trimble, 2011).

Between the 1995 and 2300 fiscal years one billion dollars per year was authorized in appropriations for the drinking water state revolving fund and the EPA was instructed to reserve 0.3% in each annual appropriation. 10 million dollars was to be directed towards research on drinking water contaminants and the potential adverse health effects. A further 2 million dollars is to be used along the lines of monitoring and regulated contaminants without 2% for technical assistance. Alaskan native villages and Native American tribes are also entitled to 1.5% of the funds each year in grants.

The state revolving fund provisions were modified by The Water Infrastructure Improvements for the Nation Act in the sense that amendments in the SDWA increased the portion of grants that the state can use for the purposes of program administration costs. Furthermore, the state revolving fund with some exceptions are not allowed to be used by States unless the water system projects that they are funding use steel and iron products that are produced within the United States (Tiemann, 2011)

4.10.12. Drinking-Water Security

We have already discussed the public health security advertisement preparedness and response act of 2002 that was passed by Congress earlier and as much as it discussed and address the wide range of security issues it made amendments to the SDWA to address any threats posed to drinking water and some key provisions along the amendments are important to mention.

The most apparent edition under the new SDWA section 1433 that was a byproduct of the public health security and bioterrorism preparedness and response act is the vulnerability terrorism attacks and other intentional acts assessments that are to be conducted on any water system that serves more than 3300 people. This is to ensure that safe and reliable drinking water supplies will still be functional and not disrupted. Community water systems are meant to certify to the EPA and submit a copy of the assessment which is meant to be undisclosed content under the Freedom of Information Act. If this information is revealed it are grounds for civil and criminal penalties (Tiemann, 2011).

The assessment should also be coupled with an emergency response plan and the requirements of the same section of the same law that is incorporated into the vulnerability assessment. The EPA is to also provide guidance especially on the smaller systems on how to conduct these assessments and compose emergency response plans and how to deal with threats. Community water systems are appropriated with 160 million dollars from the 2002-2005 fiscal years on conducting these vulnerability assessments and preparing response plants.

The Bioterrorism act also subjected the SDWA to additions and amendments that authorized the EPA Administrator to scrutinize terrorists or others would intentionally contaminate or disrupt safe water supply. Under section 1434 and 1435 of the SDWA, the EPA is mandated to investigate methods of preventing, detecting, and responding to these emergency situations and in turn have alternative safe drinking water supply if their primary sources are destroyed or disabled (Pontius, 2003).

Tampering with, attempting to tamper with or makes a threat to tamper with public water systems is also subject to civil and criminal penalties according to section 1432 of the act and even though these laws initially existed, the Bioterrorism act made amendments that increased the penalties. Maximum sentences currently stand at 20 years increased from 3 years for tampering with public water supplies and the sentences for attempting to tamper with or making threats to tamper with public water systems stand at a minimum 10 years from the previous 3 years. Fines also were increased from a maximum of 50,000 dollars to one million dollars for tampering and from 20,000 dollars to 100,00 dollars for attempts or making threats to tamper with public water systems.

The bioterrorism act also amended the SDWA to increase the grants for appropriations to not more than 35 million dollars to the EPA for technical and emergency assistance for states and public water systems for each fiscal year from 2002. Since then, fortunately, congress has not appropriated funds towards this specific purpose (Pontius, 2003).

4.10.13. Other Safe Drinking Water Act Provisions

The SDWA is very detailed in how assistance is given in the financial manner one way is in the aforementioned research, technical assistance and training in which studies and research is to be conducted towards causes, treatment, control, and prevention of any adverse health effects that is contracted through contaminated drinking water. Under section 1442 of the SDWA 15 million dollars is to be appropriated annually to technical assistance of small and disadvantaged communities and 25 million on research on the health effects of contaminants in water.

The EPA is also sanctioned to follow trends and research on novel innovative water technologies and even further to provide a technical hand in implementing and facilitating the use of new functional technology. Innovative technology grants receive an appropriation of 10 million dollars through the 2017 and 1021 fiscal years (Carter, 2006).

The monitoring inspection and record-keeping requirements according to Section 1445 of the SDWA is meant to go through the hands of the administrator to make sure regulation criteria is met. The administrator or any representative of the EPA is allowed to, after prior notification to the state in writing to inspect water suppliers to examine compliance levels. If water systems do not meet regulation standards they are subject to civil penalties.

There exists a National Drinking Water Advisory Council that is built up of 15 members that must consist of two representatives from rural systems. They are empowered by the act to advise, consult, and make recommendations to the EPA on various activities and policies that are derivative of the Act.

In the United States has Colonia that are unincorporated low-income slum areas that run along the border between the US and Mexico that are considered in the act under section 1456. Federal agencies that have jurisdiction over these areas are authorized to provide grants to Arizona, California, New Mexico and Texas in these projects but not more than half of the total costs. These Colonia also must be where access to adequate and affordable drinking water systems does not exist. For these areas appropriations of 25 million dollars were provided for the fiscal years of 1997 through to 1999 (Carter, 2006).

The SDWA is important in all these faces and more in protecting the American people from potential harmful health effects from contaminants and in implementing and enforcing legal structures and bodies to make sure compliance levels are met. The important elements to understand are the bodies or parties concerned. This is the national government, represented by congress which is the parliament version in a federation, the state government that is run like a country in itself, the public water system the EPA and the customers or citizens. The whole purpose of a government is to protect its people and in providing safe drinking water through the SDWA this is what they are trying to do. The EPA is the main body the federal government uses to particularly propagate their laws and they are the mediating body between the national and state government. The importance of the EPA for the Safe Water Drinking Act cannot be understated (Carter, 2006).

4.11. CONVENTION ON THE PROTECTION AND USE OF TRANSBOUNDARY WATERCOURSES AND INTERNATIONAL LAKES (WATER CONVENTION OR HELSINKI CONVENTION)

The Baltic Convention for the Protection of the Sea (also known as the Helsinki Convention) was originally appended by all Baltic countries along the coast in 1974 to address the increasingly severe environmental challenges posed by industrialization and different other human actions with serious effects on the marine ecosystem. The Helsinki Convention covers protecting the Baltic Sea from different kinds of terrestrial, air, and ocean pollution. It also commits signatories to take steps to protect habitats and biodiversity, as well as the sustainable usage of marine assets.

The Convention covers the entire Baltic Sea region, including inland waters and the ocean itself and the water on the bottom. Measures have also been taken to reduce land pollution in the entire catchment zone of the Baltic Sea. The first convention was appended in Helsinki on March 22, 1974. The seven parties at the time were: Finland, the German Democratic Republic (GDR), Denmark, Federal Republic of Germany (FRG), Union of Soviet Socialist Republics (Soviet Union), Poland, and Sweden.

It was further entered into effect on May 3, 1980. The Helsinki Convention got updated in the year 1992 to take account of the geopolitical variations and emergent environmental problems in the area. The Convention got expanded to 10 parties which are Denmark, European Union (EU), Finland, Estonia, Germany, Latvia, Russia, Sweden, Lithuania, and Poland.

5

Air Quality

CONTENTS

5.1. CLEAN AIR ACT

5.1.1. Introduction

The Clean Air Act (CAA) of 1970 was designed by the US Congress to protect the health of the public and its welfare from the many forms of emissions caused by the different types of air pollutants. The United States Environmental Protection Agency (US EPA) was provided authority through the promulgation of the CAA to take action in the fight against environmental pollution.

The legislation calls on the national government to put up a national air quality standard in order to minimize pollution of the air and the states to enforce these through distinct plans upon endorsement by the EAP. Air pollution regulation and national air quality standards are implemented chiefly by the states; they give out permits, check obedience and carry out checkups of the facilities, whereas EPA has the mandate to evaluate the state's actions. Among other provisions, the law calls upon the regulation of unmovable and movable sources of air pollutants and levels on harmful pollutants.

5.1.2. Background

Air pollution has been a major contributor to the health woes humanity's experience for a very long time. The United States saw these and decided to seek ways of remedying the already critical condition. In the early 1960s it came up with a legislation named the CAA (42 U.S.C. § 7401) that seeks to regulate the pollution of air on a national front. This one is the most influential law and first of the current environmental legislation. This is one of the most all round legislation in the world that deals with air quality. The U.S EAP in conjunction with the county, tribal, and local governments implements its provisions.

Prior to this act the U.S government had the Air Pollution Act of 1955 that dealt with air pollution while at the same time availing funds to the federal government for investigation into the pollution of air. The 1963 CAA was the chief national statute that dealt with the management of air pollution. This bill attained its mandate by instituting a platform within the U.S public Health Service that approved exploration into methods of observing and managing air pollution within the boundaries of the United States of America (Chay and Greenstone, 2003).

This bill has been edited several times since it became law. In 1965 it was amended to encompass the motor vehicle industry as it emerged that this industry was the biggest contributor to air pollution. Certain models made from 1968 had to meet the specifications of the Act. In 1967 the Air Quality Act came into force. This amendment sought to regulate interstate air pollution generated by industries and mobile sources. Since then several other amendments have been made to cover the following areas:

- Acidic rainfall;
- The destruction of the ozone layer;
- Toxic air pollution;
- Establish an enforcement authority for gasoline reformulation; and
- Permits for stationary sources of pollutants.

Another notable feature of this CAA is that it is amongst the chief legislation in the United States to allow its citizens to file suits in courts of law on matters to do with the pollution of the air. Several local and national governments have passed the same provisions of the act or amended it to suit their local needs and fill the gap s in the federal programs. The act has had a huge impact in combating air pollution. The graph (Figure 5.1) shows the impact of the legislation between 1990 and 2015 (Chay and Greenstone, 2003).

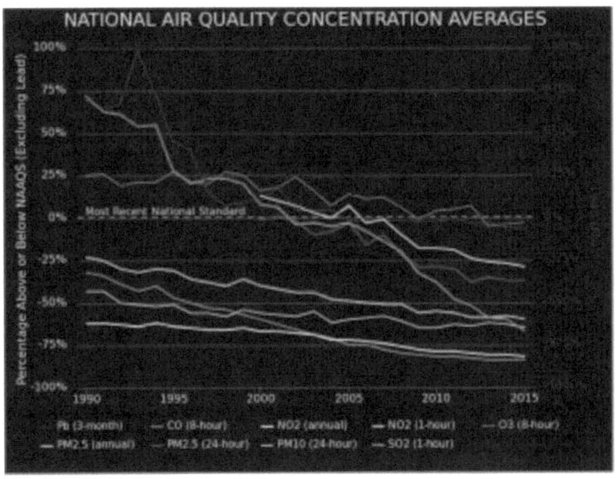

Figure 5.1: Graph showing decreases in US air pollution concentrations during 1990 to 2015.

Source: https://www.healthandenvironment.org/environmental-health/environmental-risks/global-environment/air-quality.

5.1.3. Summary of Statute

5.1.3.1. Part 1. Activities and Programs

Air quality and emissions limitations: To promote and safeguard the health of the public, the act provides for means of ensuring the nation's air quality. It aims at encouraging the prevention of regional air pollution and providing the means of controlling the pollution of the air. At both the federal and states level the act provides financial support for prevention of air pollution. There are other sub-chapters that cover the spheres of collaboration, investigation, study, teaching, and other activities related to the pollution of the air. These sub chapters also cover donations for air contamination preparation and management measures, inter-state air quality authorities and program budget minimization.

The act also classifies regions depending on their attainment of the provided regulations. There are the non-attainment regions which cannot meet the criterions for air quality. Attainment zones meet these provisions whereas unclassified parts cannot be classified using the obtainable information.

The act has established the compounds of the following as dangerous air pollutants:

- Acetaldehyde;
- Benzene;
- Chloroform;
- Phenol;
- Selenium;
- Slag fibers;
- Radioactive atoms; and
- Unregulated radioactive cadmium, polycyclic organic matter and arsenic.

Other sub-chapters cover state plan adequacy, heights of the smokestack, and the estimation of releases of oxides of nitrogen, carbon monoxide, and volatile organic compounds (VOC) from both stationary and mobile fonts. Ways of averting economic disruption and unemployment have also been covered. Land use regulations have also been dealt with here.

Ozone Protection: In 1990 and amendment was made to this chapter because of the advancement in atmospheric chemistry. The changes to the act were necessitated by the new understanding of how ozone is formed and depleted. Ozone acts as an umbrella that cushions people against the harmful UVC light and shorter wave UVB by absorbing them and letting through the less harmful UVA light. Ozone exists in the stratosphere and not the troposphere as earlier thought. Strong sunlight destroys the ozone layer and therefore it is more abundant in the poles as opposed to other regions of the earth. When oxygen comes into contact with photons from solar radiation they form ozone. Diminution in solar energy, therefore, results in a decline in ozone development in the stratosphere (Chay and Greenstone, 2003).

It has now been discovered that chlorofluorocarbons and atmospheric Freon contribute to the depletion of ozone as chlorine is a catalyst in the destruction of the ozone layer. In 1985 an ozone hole was discovered and in 1987 the Montreal Protocol called on the replacement of the chlorofluorocarbons. This necessitated the amendment of this chapter.

Prevention of Further Deterioration of Air Quality: In order to add or build key immobile sources of air pollution on has to seek a permit as stipulated by the CAA. The process of permitting these new or additional infrastructures, identified as New Source Review (NSR), relates to areas that are classified as attainment regions or unclassified ones. The permits for these areas are denoted to as Prevention of Significant Deterioration (PSD) whereas those for non-attainment regions are known as non-attainment area (NAA) permits.

The PSD program has three significant goals, namely;

- Curb growth of new non-attainment zones by guaranteeing that there is economy growth that is in agreement with the prevailing clean air.
- Ensure that the health of the public and its welfare is protected from any hostile effects.
- Protect and preserve quality of the air in areas of special interests such as national parks, recreational sites, scenic spots and historical sites.

Plan Requirements for Non-Attainment Areas: For non-attainment zones to be classified as attainment areas the CAA needs that they submit a plan as early as possible but in five or less years depending on the magnitude of air contamination and the complexity postured by the process of attaining cleaner air. The submitted must include the following:

- Permits;

- List of pollutants;
- Contingency measures; and
- Ways and means proposed to reach qualifications requirements.

For the plan to be approved or rectified it should specify who is going to implement and enforce the changes, is it the national or federal government, or both. Once the region achieves the attainment status the making an appeal for review is possible. The reevaluation request must contain a plan for keeping the quality of air high.

5.1.3.2. Part 2. Emission Standards for Moving Sources

Motorized Vehicle Emission and Fuel Standards: This subchapter deals federal principles and grants, acts that are prohibited and arrangements to control noncompliance. Study of Releases from non-road vehicles to decide whether they contribute or cause the pollution of air is also covered in this subchapter.

Motorbikes are classified in an identical way as vehicles under the new criteria of classifying motor vehicles/ engines emissions. Other divisions deal with high altitude performance modifications, compliance fees for motor vehicles, prohibiting the manufacture of engines that require leaded gasoline and metropolitan bus standards.

The passage of this statute was not easy at that time because the automobile production claimed it could not achieve the new provisions. On the other hand, senators were concerned on the impact this could pose on the budget. These sterner standards, however, saw to the invention of the catalytic converter which does not perform well with leaded gas. Lead was therefore removed from gasoline and its negative health effects became a thing of the past.

Aircraft Emissions Standards: The quality of air over the airports is poor because of the emission of VOC. These VOCs contain formaldehyde, benzene, and butadiene's which are known to cause birth defects, skin irritation and cancer. Aircraft, ground support programs, shuttles, heating systems and passenger vehicles release hundreds of tons of emissions into the air producing smog. Big towns such as Denver, Seattle, and San Francisco therefore need a Climate Action Plan and a greenhouse gas emission account. There are also federal programs that work to offset costs of reducing these emissions. An example is the Voluntary Airport Low Emissions Program (VALE).

The act requires aircraft manufacturers and airlines to adopt the provisions agreed by the International Civil Aviation Organization (ICAO). The only shortcoming of this provision is that the agencies responsible for setting carbon dioxide emission, ICAO, and EPA, have not done so. The responsibility now lies with the Secretary of Transportation, after consulting the Administrator, to come up with the regulations and ensure that inspections are done.

Clean Fuel Vehicles: Automobiles and trucks contribute a major part in polluting the air. Nitrogen oxide, hydrocarbons, sulfur dioxide, and carbon monoxide, all of which are harmful chemicals, find their way into the air through this trucks and automobiles. When certain of these said chemicals react with sunshine they create photochemical. Ocean ph. is altered, climate is changed, birth defects, respiratory complications and cancer prevalence increases on earth because of these chemicals. Motor vehicle ownership increased in the 1990s because about 58% of families possessed more than two vehicles. This means atmospheric pollution also increased (Davidson and Norbeck, 2011).

The CAA came handy in ensuring that automobile and truck emissions were reduced by ensuring that these sources of pollutants use oil fuel that meet low emission vehicle (LEV) levels and that Clean Fuel vehicle programs were adopted. Examples of clean fuels that are available but were not fully utilized are ethanol, natural gas, liquefied petroleum gas, methanol, and electrical energy. Some incentives like the California Clean Fuels Program are sensitizing the public on the need to adopt clean fuels and its demand has expanded exponentially. This pilot program in California centers on contamination management in ozone non-attainment regions. Light duty Lorries and light-duty automobiles in California have to meet the regulations. To ensure this is achieved clean alternative fuels are sufficiently distributed in the state. The manufacture of clean fuel vehicles is not mandates but it is part of this pilot program in California (Davidson and Norbeck, 2011).

5.1.3.3. Part 3. General Provisions

Before 1990, EPA was called upon by law to offer a list of Hazardous Air Pollutants and a health-based standards for each of the listed pollutants. 187 pollutants were listed as well as the origin from which they emerged from. EPA was given 10 years to develop technology-based release standards. Under these general provisions, EPA is mandated to assess any lingering risks as well as set and enact new standards with the public health protection in mind.

Citizens are at liberty to file lawsuits in order to get compliance with emission standards set by state or EPA, unless there is one already in enforcement pursued by a state agency or EPA.

5.1.3.4. Part 4. Noise Pollution

This subchapter was in existence prior to the CAA but with the coming of the act, it became classified as Title IV. An amendment was enacted in 1990 and this title and renamed Title IV-A.

This title brought with it the Office of Noise Abatement and Control to combat sound pollution in metropolitan areas, to reduce noise linked influences on physiological and psychological effects on people, wildlife, and property. The department was also tasked with running trials on the effects of noise. Title IV-A: acid Deposition Control: after the 1990 amendment this title was added to address issues pertaining to acid rain. Acidic rainfall is caused by nitrogen oxides and sulfur dioxide which are emitted from electric power plants that use fossil fuels, and industrial bases. The amendment provided industries more options on controlling toxic waste such as converting to low-sulfur coal and adding gadgets that regulate injurious emissions. In extreme circumstances industries had to be shut down to avoid further pollution of the atmosphere (The United States Environmental Protection Agency, 2007).

To achieve its mandate of reducing sulfur dioxide emissions the article had a two-pronged approached. First, more than 100 electric producing plants higher than 100 megawatts had to reduce its SO_2 emission by 3.5 million tons before January 1995. Secondly, those with a production margin more than 75 megawatts were given a deadline of January 2000.

5.1.3.5. Part 5. Permits

A national operating permit program was initiated by the 1990 amendments. This permit package covered thousands of big commercial and industrial bases of pollutants. Big businesses were required to deal with the issue of impurities released into the atmosphere, measure its size and implement a program of reducing them. These companies were further required to make periodical reports on their progress in taming air pollution. These amendments consolidated the requirements of one facility into a solitary document (The United States Environmental Protection Agency, 2007).

In non-attainment regions permits were required for those facilities that discharge 10, 25 or 50 tons annually of VOC subject to the status of the regions non-attainment. These permits are supplied by local and state authorities. In the event that the state fails to monitor adequately the adherence of the requirements, Environmental Protection Agency (EPA) may take over. If the public wishes to see the permits, it's at liberty to contact the EPA. After every five years the permits have to be renewed (The United States Environmental Protection Agency, 2007).

5.1.3.6. Part 6. Stratospheric Ozone Protection

Title VI was mandated by the 1990 amendment to regulate the use and manufacture of chemicals that damage the ozone layer in the stratosphere of the earth. The ozone layer prevents the harmful ultraviolet rays of the sun from reaching us. Scientists have linked these rays to skin cancer and cataracts.

Chemicals that harm the earth's ozone layer have been classified as either Class I or class II. Class I are those chemicals that have a ozone-depleting potential (ODP) (HL) of 0.2 or greater. These are mostly chlorofluorocarbons. Class II is those that are known, or suspected, to the destruction of the stratosphere. Both classes of chemicals that are destructive to the ozone layer have a timeline for removal:

- Class I chemicals, not more than seven years after their addition to the list; and
- Class II ten years and no more after being added to the list.

This title established means of inhibiting the entry of the above-mentioned elements from entering the stratosphere. Reutilizing and suitable disposal of these compounds were some of the proposals. Finding alternatives that are less harmful or have no harm at all was another proposal. EPA has a program dubbed Significant New Alternatives Policy (SNAP) that assesses and controls the substitutes for ozone-destroying chemicals which are being phased out under the CAA. In 1987 over 190 countries in the world engaged the Montreal Protocol, all agreed to work together to limit and eradicate the use of chemicals that eat up the ozone blanket above (The United States Environmental Protection Agency, 2007).

5.1.4. History of Controlling Air Pollution

The first legislation addressing air pollution was passed by the upper house of the United States in the year 1955. This law was called Air Pollution Control Act. It delivered funds to the United States Public Health service but did not have policies that regulate air pollution. The CAA of 1963 remedied this by a study and supervisory program in the health sector. The 1963 act allowed the establishment of standards of emissions for immobile sources of air pollution but not mobile source. 1967 came with the Air Quality Act that authorized execution of interstate air pollution standards and allowed convenient observing studies and inspections of fixed source of air pollution.

1970 came with the CAA Extension amendment that expanded the federal mandate in dealing with air pollution. Comprehensive regulations were now required by the federal government for both traveling and fixed sources of air pollution. This amendment came with the following supervisory programs:

- National Ambient Air Quality Standards (NAAQS) program to develop criteria for carbon monoxide, sulfur dioxide, nitrogen dioxide, hydrocarbons, particulate matter and photochemical oxides;
- National Emissions Standards for Hazardous Air Pollutants (NESHAPs);
- State Implementation Plans (SIPs); and
- New Source Performance standards (NSPS).

Senator Edmund Muskie played a pivotal role in drafting this amendment and that is the reasons it is referred to as the "Muskie Act." Implementing these new strict provisions of the 1970 was not an easy job. EAP administrator of that time, William Ruckelshaus, had to 60% of his time on the car industry. The automobile industry was required by the new provisions to reduce its emissions by a staggering 90%. The previous existing weak laws had given the automobile industry a free hand in running its affairs (Shouse and Lattanzio, 2020).

New provisions were added to the CAA in 1977 that included PSD for areas with NAAQS and new necessities for NAAs.

Amendments of 1990 to the CAA added supervisory requirements for management of acid rain and issuance of licenses for stationary sources of air pollutants. The aim of these new provisions was to reduce sulfur dioxide emissions which incorporated a cap- and trade package. This program enabled power generating establishments to meet the requirements of the law, unlike the provisions of the Act. The amendment expanded the NESHAP program with a list of 189 dangerous air impurities to be controlled within hundreds of origin classes. The NAAQS package was further stretched to include new requirements for fortification of the stratospheric ozone layer, increasing execution authority and expanding the study programs (Shouse and Lattanzio, 2020).

5.1.5. History of Clean Air Act

Figure 5.2: President Lyndon B. Johnson signing the 1967 air quality act in the east room of the white house, November 21, 1967.

Source: https://www.wikiwand.com/en/Clean_Air_Act_of_1963.

The 1990 Clean Air Act Amendments (1990 CAAA) contains the legitimate authority vested on national programs pertaining air contamination control. The amendments expanded the requirements of the CAA, referred to as "the Act." The first state law regarding air contamination, known as the Air Pollution Act, of 1955 provided $3 million per annum to the U.S public Health Service for a period of five years. These funds were meant for air contamination study, air contamination mitigation investigation and practical training help to the states. In 1959 the act was protracted for a period of four years with an increment in its aid to $5 million per year. The act was further modified in 1960 and 1962 but despite all these there were no federal regulations on air pollution. These duties were delegated to the state and local authorities (The United States Environmental Protection Agency, 2007) (Figure 5.2).

The 1963 CAA was a god send regulation as it laid out regulations on the control of air contamination. The act established a national program within the U.S Health Service and allowed to study into ways of observing and managing air pollution. The Air Quality Act was decreed in 1967 in order to increase the national government activities in mitigating air pollution. These provisions allowed the federal government to conduct for the first stretch ambient observing studies and inspect fixed sources of air pollutants. It also allowed the expansion of research of air pollutant release list, ambient observing ways and methods for controlling air contamination (Table 5.1).

Table 5.1: Summary of Congressional Actions Related to Air Pollution Control

Year	Act	Citation
1955	Air Pollution Control Act	P. L. 84-159
1959	Reauthorization	P. L. 86-353
1960	Motor vehicle exhaust study	P. L. 86-493
1963	Clean Air Act	P. L. 88-206
1965	Motor Vehicle Air Pollution Control Act	P. L. 89-272, Title I
1966	Clean Air Act Amendments of 1966	P. L. 89-675
1967	Air Quality Act of 1967; National Air Emission Standards Act	P. L. 90-148
1970	Clean Air Act Amendments of 1970	P. L. 91-604
1973	Reauthorization	P. L. 93-13
1974	Energy Supply and Environmental Coordination Act of 1974	P. L. 93-319
1977	Clean Air Act Amendments of 1977	P. L. 95-95
1980	Acid Precipitation Act of 1980	P. L. 96-294, Title VII
1981	Steel Industry Compliance Extension Act of 1981	P. L. 97-23
1987	Clean Air Act 8-month Extension	P. L. 100-202
1990	Clean Air Act Amendments of 1990	P. L. 101-549
1995-96	Relatively minor laws amending the Act	P. L. 104-6, 59, 70, 260
1999	Chemical Safety Information, Site Security and Fuels Regulatory Relief Act	P. L. 106-40
2004	Amendments to §209 re small engines	P. L. 108-199, Division G, Title IV, Section 428

Source: Congressional Research Service, "Clean Air Act: A Summary of the Act and Its Major Requirements," updated May 9, 2005

5.1.5.1. Clean Air Act of 1970

The 1970 amendment of the CAA gave the federal and state authorities more teeth to regulate and limit air pollution from both fixed and movable sources of emissions. Four supervisory programs came into force with this amendment.

- State Implementation Plans (SIPs);
- National Ambient Air Quality Standards (NAAQS);
- National Emission Standards for Hazardous Air Pollutants (NESHAPs);
- New Source Performance Standards (NSPS); and

This crucial bill was enacted at the same time as the National Environmental Policy Act.

CAA amendments of 1977

In 1977 the CAA was changed with key concern on the provisions of the PSD of the quality of air in regions achieving the all-important NAAQS. It also had new regulations for bases located in non-attainment zones of NAAQS. A non-attainment zone is earthly region that satisfy one or more air quality standards of the national government. This amendment included new and major requirements for issuance of permits to ensure and monitor the implementation of the NAAQS provisions. The 1977 amendment also included the implementation of an offset trading policy that had initially been applied in Los Angeles since 1974. This policy enabled fresh sources to offset their releases by buying extra decreases from the already existing ones.

5.1.5.2. Clean Air Act Amendment of 1990

In 1990 the CAA was amended to strengthen the more the power and accountability of the federal government on matters to do with air pollution. New requirements were authorized to curb acidic rainfall and for the issuing of operating authorizations to stationary sources of air contaminants. NESHAPs were incorporated into an extended program in order to control harmful air contaminants. Also, the parameters for realization and maintenance of NAAQS were revised and expanded. Other new provisions that came into force with this 1990 amendment were those to deal with the protection of the stratospheric ozone protection, enhanced enforcement authority and enlarge research programs.

5.1.5.3. Milestones

Since its inception the CAA has undergone several amendments depending on the brutality of air contamination and the available scientific information at the time of these amendments. Some of the most notable landmarks in the development of the CAA from its embryonic stages to what it is now are as follows:

The Air Pollution Control Act of 1995z

- The ever major legislation by the national government on air pollution; and
- A financed research into the extend and sources of air contamination.

Clean Air Act of 1963

- A countrywide program to address air contamination was authorized; and
- Study into ways of minimizing air contamination was authorized.

Air Quality Act of 1967

- Research activities were expanded; and
- Authorization of enforcement procedures relating to interstate transport of contaminants.

Clean Air Act of 1970

- Increment of execution authority;
- Regulation of motor vehicle releases was authorized;
- National Ambient Air Quality Standards was recognized;
- Necessities for State Implementation Plans to realize was established;
- New Source Performance Standards for modified and new immobile sources was established; and
- The National Emissions Standards for Hazardous Air pollutants was established.

1977 Amendments of the Clean Air Act of 1970

- Requirements relating to deterrence of significant deterioration were approved; and
- Requirements concerning to NAAs were authorized.

1990 Amendments to the Clean Air Act of 1970

- Programs for acid rain management were approved;
- Controls for 189 contaminants was authorized, these included those previously regulated by the national emission standards for harmful air pollutants;
- New license requirements were put in place;
- Provisions to do with the National Ambient Air Quality Standards were modified and expanded; and
- The implementation authority was also extended and changed.

5.1.5.4. Regulations

At the beginning of this campaign against air pollution there were only six mandated criteria air contaminants, namely;

- Carbon monoxide;
- Ozone;
- Particulate matter;
- Sulfur dioxide;
- Nitrogen oxides; and
- Lead.

Table 5.2: National Ambient Air Quality Standards (as of March 22, 2016)

Pollutant	Primary or secondary	Averaging time	Standard level	Form
Carbon monoxide	Primary	8-hour	9 ppm	Not to be exceeded more than once per year
Carbon monoxide	Primary	1-hour	35 ppm	Not to be exceeded more than once per year
Lead	Primary and secondary	Rolling 3-month average	0.15 $\mu g/m^3$	Not to be exceeded

Nitrogen dioxide (NO_2)	Primary	1-hour	100 ppb	98th percentile of 1-hour daily maximum, averaged over 3 years
Nitrogen dioxide	Primary and secondary	Annual	53 ppb	Annual mean
Ozone (O_3)	Primary and secondary	8-hour	0.070 ppm	Annual fourth-highest daily maximum 8-hour concentration, averaged over 3 years
Particle pollution (PM2.5)	Primary	Annual	12.0 $\mu g/m^3$	Annual mean, averaged over 3 years
Particle pollution (PM2.5)	Secondary	Annual	15.0 $\mu g/m^3$	Annual mean, averaged over 3 years
Particle pollution (PM2.5)	Primary and secondary	24 hours	35.0 $\mu g/m^3$	98th percentile, averaged over 3 years
Particle pollution (PM10)	Primary and secondary	24 hours	150.0 $\mu g/m^3$	Not to be exceeded more than once per year on average over 3 years
Sulfur dioxide (SO_2)	Primary	1-hour	75 ppb	99th percentile of 1-hour daily maximum concentrations, averaged over 3 years
Sulfur dioxide (SO_2)	Secondary	3-hour	0.5 ppm	Not to be exceeded more than once per year

Source: U.S. Environmental Protection Agency, "NAAQS table," accessed May 10, 2016.

However, advancement in technology, testing, and monitoring have uncovered numerous other harmful air pollutants. Despite this discovery of other air pollutants that were previously thought harmless, the execution of the CAA has seen to it that there is a decrease in the number of toxins in the air and the associated health complications. The EPA believes that the 1990 CAA has prevented and will remain to prevent many negative impacts of air pollution. Table 5.2 shows the profits of CAA between 2010 and 2020 (The United States Environmental Protection Agency, 2007).

5.1.6. Health Benefits of the Clean Air Act

EPA, in the year 1997 tightened the NAAQS in regard to the level of permissible earth level ozone which make up smog and the floating particulate matter which make soot. This new development became necessary after months of community review of the new recommended standards as well as the long fierce inner debate within the administration of Clinton. This led to the most controversial and contentious environmental debate of the decade. Industry groups did not that the new regulations lightly and so they proceeded to court to challenge them. They claimed that it was an abuse of the United States constitution's non-delegation code and the case eventually landed in the United States Supreme Court. In the case, in Whitman v. American Trucking Ass'ns, Inc. the Supreme Court sustained the EPA's directives. The CAA empowers the EPA to create the national ambient air quality standards (NAAQS) mandated to establish heights of pollutants that safeguard the health of the public. As a way of managing ozone and fine particle emissions, the American Lung Association and EPA endorsed in 2011 the Cross-State Air Pollution Rule (CSAPR). The campaign claimed that this prevented 400,000 asthma cases and save two million workdays and school days that would have otherwise been wasted due to breathing complications. In the case of EAP v. EME homer City Generation, cities, and power generating firms sued EPA over the new requirements but the case was finally decided in favor of EPA.

5.1.7. Roles of the Us Federal Government and States

The 1970 amendment of the CAA calls on the states to advance SIPs on exactly how they would meet the NAAQS by 1977. The 1990 CAA is a national legislation that covers the whole country but states do much of the work to implement the Act. In exchange for funding, EPA permits separate states to choose accountability for compliance with and regulations of the CAA within their boundaries. A good scenario to illustrate the above said is when a state air contamination agency holds hearings on an application for a license by a chemical or power plant or fines those found not to be in compliance with the regulation. However, this is not obligatory and some states have chosen not to take accountability of the actions of companies located within their boundaries. In this case, EPA takes those duties.

When a state decides that it wants to take over the compliance with the CAA it must submit a SIP to EPA for endorsement. A SIP includes the means the state intends to employ in cleaning up the polluted areas. Public participation is mandatory for this to be effective. Once the SIP is approved it becomes the legal framework for implementing the clean air act (CAA). An instance of a state with a law that needs compliance with the Federal CAA through the SIP is Rhode Island. The state has delegated this obligation to the state Department of Environmental Management (RI-DEM) (Welker-Hood, Gottlieb, and Suttles, 2011).

In the execution of the CAA, the federal government recognizes the unique the position the states hold. This is because of the difference of states in terms of housing patterns, local industries, and geography and weather patterns. This means that each state requires specialized approach in solving its air pollution problems. Despite this the states are not permitted to have weaker contamination controls than the least criteria set by the federal EPA.

Each SIP must be accepted by the EPA and where a SIP is found unacceptable EPA can retain the implementation of the CAA in that state. California is an example of a state whose SIP was revoked by EPA due to its failure to meet the new provisions brought in by the amendment of the CAA of 1970 (Welker-Hood, Gottlieb and Suttles, 2011).

The federal government of the United States of America, through the EPA, helps the states through availing of money, provision of scientific research, industrial designs and expert studies to support their clean air programs.

All federal funding for transportation projects in any given urban area must be permitted by the Metropolitan Planning organization (MPO). The law does not allow states to set up standards that are higher than those of the federal standards but exempts California because of its previous woes with smog contamination in the city areas. In this regard, when California's environmental organizations sets up new standards for vehicles release, it is tabled at EPA for endorsement under this waiver. The latest endorsement for California was in the year 2009. This Californian standard was adopted by other 12 states to avoid having to advance different discharge systems in their vehicles. Last year, President Donald Trump annulled this waiver arguing that sterner releases standards have made cars too costly and that by eliminating them it will make vehicles safer. Also in this bandwagon of Donald Trump was EPA's Andrew Wheeler whose arguments indicated that while the agency respected federalism, it cannot allow one state to decree the standards for the whole nation. California's Governor Gavin Newscom saw this as Trump's partisan campaign against California and promised to sue the federal government. Other 23 states including New York City, Los Angeles, and the District of Columbia joined in the suit against the administration (Welker-Hood, Gottlieb and Suttles, 2011).

5.1.7.1. State Programs

Several states, or troubled citizens of the state, have developed their own platforms and measures to encourage clean up strategies. Some examples of these are listed below;

- California has California's Clean Air Project aimed at creating smoke-free gaming air in casinos;
- Georgia has the clean air campaign;
- Illinois has the Illinois Citizens for Clean Air and Water, this is an alliance of farmers and residents to minimize the disastrous effects of large scale livestock production methods;
- New York has the Clean air NY;
- Oklahoma's Breathe Easy became effective from 1st November 2010;
- Oregon's Indoor CAA is meant to curb air pollution by irresponsible smokers;
- Texas has the Drive Clean Across Texas; and
- Virginia has the Virginia Clean Cities, Inc.

5.1.7.2. Interstate Air Pollution

Air contamination does travel from one source to a new and this means a source of pollution in one state can travel to a different state. In large cities individuals live in one state and shop or work in a different one. Air contamination from trucks and cars can therefore travel interstate. The 1990 amendment of the CAA provided for inter-state commissions on the control of air pollution, which developed regional programs for cleaning up air contamination. Under Title IV of the CAA is the Acidic Rainfall Program that allows releases trading to lessen the general budget of managing releases of sulfur dioxide.

5.1.7.3. Leak Detection and Repair

The CAA calls on manufacturing facilities to set up a Leak Detection and Repair (LDAR) package to audit and monitor a facilities emission of VOC. This strategy is aimed at identifying and repairing of constituents such as pumps, valves, flanges, compressors, connectors, and any other component that might be dripping. The main sources of quick VOC emissions are these components.

A portable vapor analyzer is used manually for testing; it reads in parts per million or ppm. In order to monitor frequency and the leak threshold, factors such as the compound tested and the substance running through the line are put into consideration. Moving parts such as pumps, agitators, and pumps are checked regularly as compared to non-moving parts such as screwed connectors and flanges. Once a leak has been detected, the regulations provide a specific time frame for repair. Most facilities are given five days for the first repair and not more than 15 days for a complete repair. Where the repair of a component necessitates an entire shut down of the facility, allowances for adjourning the repair past the stipulated time is made.

5.1.7.4. Application to Greenhouse Gas Emissions

On January 2, 2011 the EPA started regulating greenhouse gases (GHGs) from both movable and immobile sources of air contamination for the very first time under the CAA having established its auto releases standards in 2010. Pursuant to section 202 of the CAA the provisions for mobile sources of air pollution was established. Under the directive of Part C of Title I of the CAA GHGs from stationary sources of air pollutants are controlled. The EPA's (EPA) auto emissions levels for greenhouse gas releases issued in 2010 and 2012 were meant to reduce releases from targeted vehicles by half, double the fuel economy of traveler vehicles and light-duty trucks by 2025 and save over $4 million drums of oil and $1.7 trillion consumer money. The EPA has also suggested a two-pronged approach to decrease greenhouse gas releases for average and heavy duty trucks and buses.

5.1.7.5. Clean Air Act and Environmental Justice

The CAA helps reduce exposure of people of color and low earnings groups to air contamination by promoting the reduction of air pollution. Dr. Marie Lynn Miranda, an environmental scholar, says that African Americans are over characterized in zones with the poorest air quality. In the United States of America, the most densely populated regions are inhabited by low-income earners and the minority communities. This scenario heightens the health complications of these inhabitants. Exposure to high intensities of air contamination is associated to health complications such as asthma, premature death, cancer, and newborn mortality, which highly effect negatively on low income earners and communities of color. The CAA significantly reduces air pollution which directly results in the decline of health problems faced by communities of color and low income earners. This, therefore, brings justice to these disadvantaged communities by ensuring that despite the economic standards or skin color we all enjoy a pollutant free environment.

5.1.7.6. Clean Air Act Violations

Violators of the CAA are brought to book by the EPA according to the severity of their violation. Parties or companies that go against the provisions of the CAA are imposed with monetary penalties. The EPA has the authority to fine public officers $37, 500 per day (with a maximum of 8 days) except otherwise provided by EPA. Amongst the major cases involving the violation of the CAA are:

- The 2015 Volkswagen emissions scandal.
- Violation of diesel engine emission standards by caterpillar and five other manufacturers (consent decree, July 1995).
- Hyundai and Kia violated the CAA which resulted in a penalty of $100 million to the United States and to California Air Resources Board.

5.1.7.7. Effects of the Clean Air Act

The CAA has significantly condensed the level of air pollution in the United States of America by giving more legal teeth to both the federal and state environmental agencies. In 2017 a study showed that there has been a 10%decrease of contamination in counties that went beyond the threshold set by the 1970 amendment of the CAA in the 3 years after the guideline was enforced. The study further established that this regulatory prompted drop in air contamination has directly increased the working hours and earnings of the affected workers by one percent in yearly wages.

The writers of the study estimate that lifetime collective income gain for the affected individual is $4,300 in the present value terms. Furthermore, the study shows that with the improved air quality in the United States, 205,000 early deaths and millions of other breathing illnesses were averted. This is estimated to have had an impact on the economy worth $50 trillion in savings as compared to the $523 billion invested in implementing the CAA standards.

Since the 1970 amendment of the CAA, movable sources of air contamination such as trains, vehicles, boats, and motorcycles have become 99%cleaner for contaminants such as carbon monoxide, hydrocarbons, particle emissions and nitrogen oxides. The acceptable emissions for volatile organic chemicals, nitrogen oxides, carbon monoxide and lead from personal cars have been condensed by more than 90%ensuing in a national reduction of these contaminants notwithstanding the more than 400%increment in miles driven annually.

Beginning from the 1980s, 25% of earth level ozone has been reduced, mercury releases reduced by 80%and ever since the change from leaded to non-leaded gas, 90% of atmospheric contamination has been condensed.

An investigation in 2018 showed that the CAA has caused a 60%waning in air contamination by manufacturing industries between the year 1990 and 2008.

5.1.8. Future Challenges on Clean Air Act

The global phenomenon of climate change is posing a huge test to the management of the normal types of pollutants in the United States of America because of the warm, dry summers settings that can result in air inactivity episodes. Wildfires might also increase due to prolonged droughts resulting in regional high levels of air particles.

Some cities in the United States of America, as of 2017, still do not meet all NAAQS. There is there the likelihood of thousands of untimely deaths caused by fine particle contamination and earth ozone level air contamination. Another big challenge is that air pollution is not confined to one nation. Pollution from upwind countries often finds its way to the United States making it even the harder to meet the provisions of the CAA. On the other hand, air traveling downwind of the U.S might contain contaminants already in it. Tackling these challenges require international cooperation in reducing the pollutants in the originating country.

5.1.9. Proponents of the Clean Air Act

Any legislation that deals with the environment often raises opposing views from the concerned stakeholders and the government. There are those who support the law and those who criticize it. Below are some of the proponents of the CAA.

- The Union of Concerned Scientists in an article titled The CAA argues that as a direct outcome of the Act there has been a 25% drop in smog, sulfur dioxide, and nitrogen dioxide reduced by 71% and 46% correspondingly, and decreases in lead releases by 92%. Reduction in lead emission was occasioned by the CAA's provision that refiners ought to reduce the lead levels in gasoline. All the above-mentioned reductions have occurred between 1980 and 2016. The article furthermore leads the CAA for its standards that require plants and other industries to put in place new anti-pollution technology to manage air pollution.

- The center for American Progress in a November 2015 article titled 'protecting public Health and Growing the Economy: 25 years of the 1990 CAA amendments' argued that the Act has not hindered economic development or restricted energy consumption. On the contrary, there has been a noteworthy reduction of destructive pollutants over the said period despite an increase in population, vehicle miles, and the gross domestic product. Pollution s of lead, nitrous oxides and sulfur oxides has fallen by 80, 51, and 79%respectively.
- Susan Brooks, the director of U.S Climate policy and Analysis at the Environmental Defense Fund, in a post titled 'The CAA: Good for Our health and Our Economy' cited a research authored by the EPA that showed that the CAA has generated $1.3 trillion in wellbeing and eco-friendly benefits with an investment of about $50 billion. This research by EPA further indicates that by the year 2020 the Act would have assisted avert 230,000 untimely deaths and produced $2 trillion fiscal benefits. The CAA has therefore helped in preventing premature deaths in the United States and saving trillions of dollars.

5.1.10. Criticism of the Clean Air Act

For more than 40 years the EPA, has worked relentlessly to eliminate harmful air pollutants in accordance with the provisions of the 1970 amendment of the CAA. These noble efforts have however been hampered by a flaw in the statute: the exemption of existing industrial facilities. The most notable of those exemptions are the coal-fired power industries. These plants ought to be exempted from the federal limits on some the most common yet hazardous types of pollutants. This exemption has therefore continued to adversely affect the health of several people.

Secondly, the pairing of standards for both new sources and existing ones has created problems. Old sources are allowed to operate with lax standards or no standards at all whereas the new ones are required to meet restrictive standards. This has encouraged old plants to operate longer than otherwise they would have if the same standards were applied across the board. On the other hand, new plants are discouraged by these restrictive measures from operating.

In 2007, Joel M. Schwartz and Steven F. Hayward published a study entitled 'Air Quality in America' carried out by the American Enterprise Institute. They concluded that prior to the CAA of 1970 air contamination was already declining in the United States. They mentioned the city of Pittsburgh, which in the years between 1920 and 1940 reduced its airborne particulate substance by a percentage of 50 and another 50%between 1940s and 1970s. Also cited on the publication were the Midwestern and Eastern industrial cities that had attained a big success in the reduction of particulate levels through the initial and middle years of the twentieth century. Statistics from the city of Los Angeles show a sound level of reduction of smog through the 1960s and those from New York indicate a 58%reduction in sulfur dioxide levels from 1963 and 1970. The taxpayers' money used to implement the CAA was therefore wasted as the nation was already recovering from air pollution.

Nick Loris, the Herbert and Joyce Morgan Research Fellow at the Heritage Foundation, while addressing the U.S House Subcommittee on Interior, Energy, and the Environment in March 2017 also rubbished the expenditure on CAA. He argued that legislation pertaining to the environment such as the CAA are not well equipped in dealing with conservational issues and at the same time are costly with little or no benefit to the environment. He said "The EPA has used ever-expanding authority to implement stringent regulations with increasingly high compliance costs and diminishing marginal environmental returns,"

The executive director of the Property and Environment Research Center, Reed Watson, in a December 2016 report titled 'Giving Credit Where is Due,' argued that the reduction of air pollution between 1970 and 2015 were as a result of local government decrees and economic development rather than the CAA. He further pointed out that air contamination had started decreasing between 1962 and 1970 (when the CAA was altered) and that after the acts passage air quality improvements slowed since, "federally mandated reductions often failed to account for local conditions, creating expensive and often ineffective one-size-fits-none approaches to clearing the air." he said.

5.1.11. Other Views on Clean Air Act

Others feel that the CAA has significantly improved the excellence of air in the United States but still it has some flaws on its priorities. Writers of a 2004 report titled 'Air Quality Management in the United States' and published by the National Research Council claims that the CAA will undoubtedly remain to have economic paybacks. To substantiate its allegations the authors claimed that the EPA's execution was at times engrossed on course rather than outcomes. The report observed that the procedure states use to attain the NAAQS was a legalistic and often exasperating which focused solely on compliance rather than the end results. According to the authors, these discouraged states from innovating and experimenting with new ways of controlling air pollution. They clinched that the states and localities were weighed down provided their narrow human and monetary resources to conform with the requirements of the federal EPA. States and local governments therefore had to set aside resources to meet these stringent provisions.

5.2. THE MONTREAL PROTOCOL

The Montreal Protocol, completed in 1987, refers to a global agreement aimed at protecting the stratospheric ozone level by phasing out production and consuming of 100 ozone-depletion substances (ODS). The stratospheric ozone level filters out dangerous ultraviolet rays, which is related to the increased numbers of cataracts and skin cancer, decline in agricultural productivity and the destruction of marine ecosystems. The United States passed the Montreal Protocol back in 1988 and joined together with the four successive amendments.

The Montreal Protocol has proved to be an inventive and successful legislation, and is the original treaty universally ratified by all nations in the world. Using participation from all over the world, the Montreal Protocol provided a clear signal to the universal market and has put the ozone layer previously at risk on the road to repair. By the end of this century, in the US alone, the full application of Montreal Protocol is anticipated to avoid nearly 280 million skin cancer cases, about 1.6 million skin-cancer fatalities, and 45 million cataract cases thus bringing greater benefits globally.

This treaty gradually phases out the usage and production of different ozone-depleting substances, and sets different timetables for developed as well as developing countries. According to the treaty, all parties have unique responsibilities related to the elimination of different types of ozone-depleting substances, control of the trade in ozone-depleting substances, annual reports of data, and national licensing systems for controlling the import and export of ozone-depletion substances, and other factors. Developing and developed nations have equal though differentiated responsibilities, but more significantly, both groups of nations have binding, time-bound goals and measurable commitments. (UNEP, 2020) Available at: https://www.unenvironment.org/ozonaction/who-we-are/about-montreal-protocol

6

Endangered Species of Wild Fauna and Flora

CONTENTS

6.1. INTRODUCTION

Endangered species can be defined as a species that stands a very high chance of becoming extinct in the projected future. This can be on a local level, continental level, national level or it can be worldwide. There can be several factors that can lead to the declaration of a species being endangered, for example, loss of their habitat, being poached for their hides, skins or body parts, as well as invasion by another species. There are various international bodies and agencies that are mandated with the listing of the endangered species as well as the conservation status of various animals across various areas and globally as well. Many countries in the world have developed and implemented laws for the protection and conservation of these species. These laws bar human activity in the conservation areas and also practice of harmful human activity in places where animals are endangered. Some of these activities are: hunting, land development restrictions, and creation of protection zones. There are also some endangered species that are seen as the bulls eye of major rescue and conservation efforts for instance captive breeding as well as habitat restrictions (Figure 6.1).

Figure 6.1: An image of a Siberian tiger, one of the most endangers species in the world.

Source: https://storymaps.arcgis.com/stories/fe7589e228e7497292135d57ed64e5b6.

There can be natural causes of forces that lead to the endangerment of species all over the globe. The natural causes such as climate shits and changes as well as geological changes, they are considered to be relatively lower than human factors and activities. This puts an enormous strain on the animal or plant species population eventually leading to an imbalance in natures' functioning. Some of the causes that lead to species being regarded as endangered are for instance:

- *Habitat loss*: this is considered to be a major factor. This is because habitat loss is mainly caused by human activity. Urbanization, deforestation, construction of dams and other structures, and also agricultural ventures affect greatly the natural habitat of these species.
- *Invasive species*: this is regarded as a biotic factor, and it is all about the eco-system stability in the introduction of new species. They are likely do die-off or adapt very slowly to the new ecosystem though at times they are able to exploit the new ecosystem that they are introduced to and threaten the existing species of that particular ecosystem.
- *Pollution of the environment*: various forms of pollutions have different effects of the sustainability of a species in a given area. There can be the common forms such as noise, light, water, and air pollutions. These pollutions are seen on an industrial level, but humanity fails to look at the smaller forms. For example, the mosquito repellants that are aerosol sprays, have an effect of the reproductive rates of birds as shown by research. Therefore, an area that has a lot of mosquito breeding is likely to have a low population of birds in the area which will affect the ecosystem.

Due to the extensive hand of man in the population of species has led to the development and the implementation of laws that regulate the human activities as well as the extent of these activities. This paper seeks to look at the Endangered Species Act under environmental laws, its history as well as the processes that accompany the listing of these species and the effectiveness of this law (Foreman, 2002) (Figure 6.2).

Figure 6.2: Tiger and leopard skins rescued at customs in China.

Source: https://www.dw.com/en/tigers-prowl-texas-backyards/a-44347435.

The Endangered species act was implemented and made effective in 1973 and it became the primary law in the United States of America and its peripheral territories, signed into law by president Richard Nickson. This law seeks to offer protection to imperiled species also known as endangered species such as fish, plants, and wildlife from extinction. This was referred to as "the most comprehensive legislation" by the U.S. Supreme Court to be set up by any country for the protection of its species. There are two main objectives of the Endangered Species Act;

- Prevention of extinction; and
- To recover species until the point where the protection by law is not necessary.

There are agencies set up by the united states government housed by different arm of the same government to oversee the processes of this act such as The U.S Fish and Wildlife Service that is under the Interior Department, The National Oceanic and Atmospheric Administration found under the Department of Commerce. All the above-named bodies handle the management and the conservation of fish, wildlife as a resource as well as their habitat with an inclusion that they are also responsible for most especially the endangered species. In the past, there were no regulations that protected the population of species and the ecosystems that they depend on. Activities such as hunting animals like tigers and bears for among others for their hides and fur, the hunting of bison to near-extinction and other human activities that cost the lives of large numbers of animals and fish had to be banned. This gave a major driving force for the call for discussions on wildlife conservation in the early 1900. The major hunting of certain species of animals, birds, and fish was primarily for the animal markets as well as for keeping as domestic pets which raised concerns for the treatment of these species as well as the concern discussions about their habitat loss due to human encroachment. This led Congress to implement the Lacey Act of 1900 that was one of the first laws that spoke strongly with regard to regulate animal markets (Foreman, 2002).

This law also spoke against the selling of animals that had been killed illegally between one state and another state. With this law setting the pace, there were a lot more people talking about endangered species and their conservation. An example being the Migratory Bird Conservation Act that was enacted preventing hunters from hunting of right as well as the grey whale, and the Bald and Golden Eagle protection act prohibiting the hunting of the golden eagle (Figure 6.3).

Figure 6.3: *An image of the bald eagle soaring.*

Source: https://www.longislandpress.com/2018/12/02/bald-eagles-soar-over-long-island-once-again/.

In the case of the Bald eagle and the whooping crane, the treaties and the protection laws did not seem to hold due to these laws being ignored, many populations of these birds and other animal species still kept declining. There was an estimated 16 whooping crane in the wild by the year 1941. At the time of 1963, there were only 487 nesting pairs left for the bald eagle which id the United States national symbol due to shooting and habitat loss. There was a great need of funding for proper implementations and monitoring of these species as well as congressional authority. Due to the need for funding among other crucial supporting factors, Congress went ahead to pass the Endangered Species Preservation Act in the year 1966 which was set in place for the restoration and the protection of the native fish and wildlife species and a further mandate of the department of Interior to acquire and give land for the conservation of these species as conservation areas. This saw to the first ever listing of species that were considered endangered by March of 1967, which was inclusive of:

- 14 mammals.
- 22 fish.
- 6 reptiles.
- 36 birds.
- 6 amphibians.

There was also an inclusion of species like the grizzly bear, the bald eagle, the American alligator, and the Florida manatee. A key item to be noted is that at the time, the list was only made up of vertebrates due to the fact that the act in conjunction with the Department of interior's definition of endangered species was only wildlife and fish (Foreman, 2002).

CITES (Convention on International Trade in Endangered Species of Wild Fauna and Flora) is a global agreement between governments. The purpose is ensuring that universal trade in wild animal and plant specimens don't affect their survival. Today, extensive information about the endangered condition of many important species (such as tigers and elephants) may make the need for these conventions seem obvious. However, when the idea of the CITES of Wild Fauna and Flora was first developed in the 1960s, the universal discussion on the protection of the wildlife trade was comparatively new.

In hindsight, CITES' needs are obvious. Each year, the global wildlife trade is believed to be worth several billions of dollars, including several millions of animal and plant specimens. Trade is diverse, from live animals and plants to a variety of wild animal products derived from it, such as food, exotic leather products, wooden musical products, wood, travel antiques and medicine. The extent of exploitation of certain fauna and flora species is considerably high, and their trade and other factors (such as habitat loss) can seriously consume their populations and further make some species become endangered.

6.2. ENDANGERED SPECIES CONSERVATION ACT OF 1969

6.2.1. Introduction

A few years later, 1969 saw the amendment of the initial US Endangered Species Preservation Act of the year 1966 to the Endangered Species Conservation Act of 1969. This covered and upheld the initial goals set by the previous Act which was to protect and conserve the endangered species as well as a very valuable addition in the new Act. This was the listing of species that were facing near extinction worldwide, as well as the inclusion of the endangered species listed in the 1966 Act. It expanded its' reach in comparison to the act of 1966 that focused on wild birds and game, this Act of 1969 covered those and the addition of protection for mollusks and crustaceans. The repercussions for going against these Act in the form of poaching or illegal importation and sale of the listed species was hiked to result to fines of up to $10,000 or a jail term of one year. This Act also pushed for the creation of conventions on an international basis and the making of treaties and agreements for the conservation and protection of these species that were endangered. February of 1973 saw the production of the Convention on International Trade of Endangered Species of Wild fauna and flora that came from the formation of the comprehensive multilateral treaty at a meeting held in Washington D.C. this Act of 1969 provided a guideline for the Endangered Species Act of 1973 that worked on the sole information given by scientific and commercial date that set the bar to determine whether or not a species was categorized as endangered and well on its way to extinction.

President Richard Nixon spoke out and stated that the current implementations that were to conserve and protect spices was to be considered inadequate in 1972. The Endangered Species Act of 1973 is considered to be one of the most important environmental laws and it is the major landmark for conservation law both in the United States of America but in the world as a strong protection law for endangered species. There was however a lot of biodiversity protection laws and science affiliated with this conservation law with regard to these species. To add to that, habitat loss was considered to be a serious threat to millions of species, those endangered and not. Factors like climate change, invasive species, changes in land use and land overexploitation stood and still do stand as threats to the survival of these species and the eco-system that they depend on for a living (Doub, 2016). The content of the Endangered Species Act consists of 18 sections that cover the legal aspects, international factors as well as various aspects of the coverage of the conservation law in regard to different species of fish, animals, plants, and other endangered species in the world and the United States America. The sections of the Endangered Species Act are:

- Section 1 – Preamble.
- Section 2 – Findings, purposes, and policy.
- Section 3 – Definitions.
- Section 4 – Determination of endangered species and threatened species.
- Section 5 – Land acquisitions.
- Section 6 – Cooperation with the states.
- Section 7 – Interagency cooperation.
- Section 8 – International cooperation.
- Section 8a – Convention implementation.
- Section 9 – Prohibited Acts.
- Section 10 – Exceptions.
- Section 11 – Penalties and enforcements.
- Section 12 – Endangered plants.
- Section 13 – Conforming Amendments.
- Section 14 – Repealer.
- Section 15 – Authorization of appropriations.
- Section 16 – Effective date.
- Section 17 – Marine Mammal Protection Act of 1972.
- Section 18 – Annual cost analysis by the Fish and Wildlife Service.

Figure 6.4: A rescued rhino kept at the Hoedspruit endangered species center.

Source: https://jumbari.com/hoedspruit-endangered-species-centre/.

Section 1 – (preamble): This first section mainly looks at giving and introducing the Title as well as the exact purpose of the Endangered Species Act which is basically to give protection for threatened and endangered species of plants, animals, and fish from extinction (Doub, 2016) (Figure 6.4).

Section 2 – Findings, Purposes, and Policy: This section goes into providing a basic background of the act and defining the findings that congress came to at the time of the enactment of the Endangered Species Act. It goes on to look at the history of the Endangered Species Act as well as history of congress with regard to this act and the intentions of the Acts enactment. It looks at the policies that were put in place as a follow up of the findings and the taking action to carry out the purpose of the Act (Doub, 2016).

Section 3 – Definitions: This section deals with the definitions of terms throughout the document. Terms like endangered species, fish, and wildlife, critical habitat, convention, permit or rather license application, conserve, among 15 other terms are all defined in this section in the focus and deliberate step to ensure that this Act is understood and the concepts are well comprehended by those that it concerns which is the united states and its peripheral Nations/extensions (Doub, 2016).

Section 4 – Determination of Endangered Species and Threatened Species: This section gives us an in depth look at the structural formation and functioning of the Endangered Species Act as it stands around the purpose of protecting, rescue, and restoration of the species that are classified and considered endangered or threatened into extinction. It also goes into offering information on the listing process of these species as well as the eventual removal form the list in the event the mission is accomplished, and the population of the species is well above self-sustainability and the discouragement of harmful human activity to the species.

Section 5 – Land Acquisitions: With the conservation and the rescue of these endangered and threatened species, there is a need for space where possible breeding and population boosting is worked on. This section focuses on exactly that. The acquisition of land and the various procedures that are in regard to that. Land can be bought from private landowners or it can be bought or allocated from governmental owners of these pieces of land. The land is majorly used for habitat creation for the species in hope to improve their population numbers enough to be out of danger.

Section 6 – Cooperation with the States: The united states government set a mission to encourage the conservation of these species that were listed by the Endangered Species Act. This is what this section in the Act focuses on. The building of relationships between states in the aim of moving in the same direction of protecting the endangered species. It also discusses the cooperation between the federal agencies and the states as a key element to the success of the purpose of the Endangered Species Act found in cooperative agreements, management agreements, the fund allocations between the federal agencies and the states. There is also coverage on the conflicts the laws that have been set by the federal government and those laws set by the states.

Section 7 – Interagency Cooperation: Due to several conflicts that arose in different federal agencies there was a need to address the way forward in handling any issues that would come up with regard to the Endangered Species Act. There was the formation of the Endangered Species Committee for the primary function of being the arbitrator in these meeting that were set up to clear and handle the issues at hand. This section lists the formation of this committee as well as how these conflicts are to be dealt with maintaining focus on the goals of the Endangered Species Act.

Section 8 – International Cooperation: On the enactment of the Endangered Species Act, the key focus was to implement it in the United States of America as well as the extending this aim to international communities. This section covers the relationships created between the American Federal Government and foreign governments in the discussions that lead to treaties being created and signed into law across the nations that are in discussion with the United States. There is also an inclusion of the employees that will handle the implementation of the Endangered Species Act in these different nations as well as the follow-up investigations by the responsible bodies as to the upholding of the key purpose of the Endangered Species Act. There is the coverage of the financial agreements including assistance for those countries that cannot handle the costs of this enactment and the encouragement of several foreign programs all driven by the sole purpose of the Endangered Species Act.

Section 8a – Convention Implementation: With the establishment of relationships between the Federal Government and foreign governments, the funding matters aligned and ironed out, the need for identification of special procedures as the key element for the implementation of the International Cooperation. This section covers the actual specific procedures and the coming up of international conventions for the conversations that would encourage the drive for the Endangered Species Act purpose and key goals which was the conservation and the restoration of the listed endangered species.

Section 9 – Prohibited Acts: This is one of the most important sections in the Endangered Species Act. The Endangered Species Act stands as law and not adhering to this law was and is still considered a violation of the law. This section talks clearly about the violation of the Endangered Species Act and it goes ahead to specify the various types of violations that can be committed by man. Some of the violations are:

- The holding captive of species that are considered to be endangered or are listed in the Endangered Species Act or putting these species in a controlled environment.
- The violation of any Convention that has been enacted and implemented under the Endangered Species Act.
- The importation and the exportation of these species in the protection of special species such as the African Elephant and the Rhino for their horns as ivory.

- Regulations that cover importation and the exportation of these species and their produce.
- Reports that cover the importation and the exportation of fish and wildlife.
- The designation of ports.

Section 10 – exceptions: There are exceptions that are set aside by the Endangered Species Act and this section covers in detail these exceptions to preventions that are introduced by the Endangered Species Act. The covered exceptions are: hardship exemptions, exceptions that are given to a specific group of native people, permits for experimental populations as well as an array of other different exemptions under the Act.

Section 11 – Penalties and Enforcements: This section of the Endangered Species Act brings into play the penalties that individuals or groups that will violate the Act at both civil and criminal violations. The section goes on further to elaborate on the district court jurisdictions, the rewards and specific incidental expenses, the enforcement measures, the regulations, and an inclusion that covers private citizen suits and the specific procedures that maintain them under the Endangered Species Act.

Section 12 – Endangered Plants: This section introduced the coverage of endangered plant species under the protection of the Endangered Species Act.

Section 13 – conforming amendments: This section covers the amendments of related Acts for example;

- The Migratory Bird Conservation Act of 1953.
- The Land and Water Conservation Fund Act.
- The Marine Mammal Protection Act of 1972.
- The Federal Environmental Pesticide Control Act of 1972.

Section 14 – repealer: This is a brief section covering previous repeals of different provisions of the Endangered Species Act.

Section 15 – Authorization of Appropriations: The contents of this section is the authorization in relation to the initial funding of the operation of the Endangered Species Act inclusive of subsequent re-authorization of the funding process through the year 1922. It also covers exemptions from the Endangered Species Act and the convention implementations.

Section 16 – Effective Date: This section speaks in its entirety that - "This Act shall take effect on the date of its enactment."

Section 17 – Marine Mammal Protection act of 1972: This section covers the Marine Mammal protection stating that apart from the provisions given in the Act, there shall be no precedence taken over an additional number of restrictive provisions that cause conflict with regard to the Marine Mammal Protection Act 0f 1922.

Section 18 – Annual Cost Analysis by the Fish and Wildlife service: The statement of this section is to mandate for an accounting audit to be carried out by the Secretary of Interior through the Fish and Wildlife service that is carried out species by species basis to look at the expenditure that was covered by the federal government in its funding under the ESA in the goal to uphold the purpose of preservation of the endangered species that are listed in the Act. This is important for the evaluation of the spending in the proceeding into the new financial year (Nazzaro, 2010).

Over all, the coverage of the Act is to outline the basic functional rules and it looks at the violations I order to have a greater understanding of the Endangered Species Act is understood and is taken into consideration by the citizens of the United States of America as well as the international nations that have opened their mind to the conservation and protection of the endangered species for the benefit of the environment as well as the ecosystem they depend on and for the survival of humanity due to the importance of these species whether plants or fish, birds or animals in the wild as game, they play an important role in our eco-systems (Nazzaro, 2010).

6.2.2. Listing Process of Endangered Species Act

As earlier mentioned, the listing process is covered in Section 4 of the Endangered Species Act and it created demarcations in the form of a five-step criterion that has to be met by a species in order to be considered for listing. The steps include,

- The evidence of danger or threat destruction, curtailment or modification posed to its habitat or range.
- The presence of extensive utilization aimed at recreational, scientific, educational gains or commercial purposes.
- The population decline of the species because of diseases or a biological factor called predation.
- The existing incompetence and inadequacy from the mechanisms set for regulation.

- Presence of other natural factors or human activity factors that interfere with the current or future existence of the species.

The species that face serious risk and threats to their existence and their well-being are presented with the highest priority and are immediately placed on an "emergency list." There are two ways of listing the species. The United States Fish and Wildlife Service could place a species on a list directly via the candidate assessment of the species program or individuals as well as organization petitions can have a request put in with the Fish and Wildlife Service for the listing of the threatened species. The petitions can be filled in with the National Marine Fisheries Service (NMFS). Species protected under the arm of the Act can be actual taxonomic species, a distinct population segment in the case of vertebrates as well as a subspecies of the original. This listing process is conducted in the same manner across the board with an exemption of the organizational or individual petitions. The procedure has a 90-day screening period (Sullins, 2001).

In the listing process, there is no consideration of economic factors, though they can be with reference to the best available commercial and scientific data presented. Looking at the amendment of The Endangered Species Act of 1982, the addition of the word "solely" which restricts the consideration apart from the biological status of the species. With this, President Ronald Reagan passed an executive order which needed the economic evaluation of all government agencies' events and actions. This was turned down by Congress and the House of Committee's stated but there was no relevance are they economic consideration with regard to species status. A similar but opposite event took place with the amendment of 1978, saw Congress making an addition of the words "…taking into consideration the economic impact…" amounting to the designation of critical habitats provision (Sullins, 2001).

There was an almost halt to the listing of new species due to the amendment of 1978 with a link being created between the procedure of listing with critical habitat designation plus the economic consideration bringing a lot of controversy within government agencies and congress. This saw a number close to 2000 species being withdrawn from the consideration process for potential listing.

Once a petition for listing is received by the relevant agencies and bodies, the two main federal agencies proceed to a subsequent stage which is the rulemaking procedure and sees each step all the way to registration in the Federal Register as they are recorded in the register which is the united states government's official journal for the rules and regulations that have been adopted and or proposed. The two steps state that;

- In the presenting of. Petition that suggest that a species is endangered, there is a 90-day screening period that begins on the day of the petition being filed for individual and organizational petitions alone. In the event there is no presentation of substantial or compelling information by the petition that calls for listing, it is declined.
- In the event that the information presented by the petition is compelling enough, the commencing of a status review is called for and it is a comprehensive assessment of the biological status of the species as well as the standing threats to their life and well-being. This is as a result of "wanted" or "not wanted" or in some cases "warranted but precluded."
- The findings of a not warranted situation, all end with the listing process being canceled.
- Findings recorded from a warranted situation mandates the agencies to publish a 12-month finding that is to be within one year of the petition date with the listing of the species as endangered or threatened.
- The "warranted but precluded" situational findings, there is an automatic back recycle down to the 12 months to an indefinite time period until there is the change of the situation to a warranted or a not warranted status is reached. This particular status is followed closely by the agencies due to its special nature. Notably, the "warranted but precluded" finding status stands as a deferral that was included as an addition in the amendment of 1982 of the Endangered Species Act.
- By the close of another year, there must be a final rule made Determining whether the species should be listed or not. This final rule has a time limit that could extend over a period of six months under the listing could be grouped with regard to similarity in geography, the habitats, the threats as well as the taxonomy.

The rate of listing per year has fluctuated over the years but started with an increase from the Ford administration which saw 47 listings which was 15 listings per year, The Carter administration which saw 126 listings that translates to 32 listings per year, the Reagan administration that saw 255 listings that translates to 32 listings per year, George H. W. Bush who's administration listed 231 that translates to 58 listings per year, The Clinton administration That recorded 521 listings translating to 65 pie year, which was succeeded by a steady decline during George W Bush's administration which saw a listing rates of 60 listings translating to 8 per year as recorded in 2008. The rate of listing is strongly dependent on the involvement of the citizens as well as mandatory punctuality meaning that as the agencies discretion presents itself as a decline and the citizen involvement presents itself at an increase this creates room for better listing rate often being more listings done. What is termed as citizen involvement is the basic working hand in the hand of the responsible bodies that uphold the Endangered Species Act's purpose with the citizens who helped in identifying the threatened or endangered species as well as increased the efficiency of the process (Shogren, 2013)?

As an extension of the listing process, the public is notified through legal notices that are found in the newspapers, notices that are communicated to state and community agencies that are around the area where the species is found geographically. In special cases for rain missions both are in conjunction with the Endangered Species Act are also likely to receive notices of a listing. If there is a request for a hearing placed within 45 days of the public notice, the hearing is considered mandatory and it is held. The sole importance of the public notice as well as the requests for comments is to present opportunity for participation of the public and this is considered very important as they are needed in the rulemaking process.

6.2.3. Listing Status of Endangered Species Act

The listing status is considered to be the placement of the species into categories depending on the status of the species in relation to extinction, threatened, endangered, and the eventual delisting of the species. The depiction represented in Figure 6.5 shows the abbreviations that are used in the status of listing of the species, and these terms and abbreviations are used by the federal agencies responsible for these processes in the federal register.

- E stands for endangered. This is a category for any species that is threatened by a significant portion of its range with an exception of species found in class, which is determined by the Secretary seen as a category of pests. (US Environmental Protection Agency, 2017) Available at: https://www.epa.gov/laws-regulations/summary-endangered-species-act

- T stands for threatened. This class is for those species that are deemed as threatened and are likely to become endangered species in the predicted future throughout all or a significant portion of its range.

Figure 6.5: listing status image depiction in accordance to the endangered species Act.

Source: https://commons.wikimedia.org/wiki/File:Status_ESA.svg.

- C stands for candidate. Species under this category are in line for consideration to be listed.
- E(S/A), T(S/A) stands for endangered or threatened as a result of similarity of appearance. The species in this category is deemed as not threatened or and angered but qualifies to be here due to its resemblance in appearance to the species that is threatened and or endangered as a result of the difficulty experienced by the agencies in trying to differentiate the listed and the unlisted species.
- XE, XN stands for experimental essential or non-essential population. This category is for any population endangered species that is released outside of its current range which is authorized by the secretary. Experimental and non-essential populations of these species are always species on public land for the purpose of consultation and as the species goes through the process of listing into private land.

6.2.4. The Survival and Recovery of Endangered Species

As stated above, critical habitat is a link between recovery goals and habitat protection that requires the identification as well as the protection of all water bodies, land, and air regions all in the effort to recover endangered species and protect them and their habitats. The consideration of food, water, light, breeding sites, seed germination and dispersals, population growth as well as the lack of disturbance all that is required and considered very essential in the determining of the exact nature and character of critical habitat. Under the Endangered Species Act of 1973, there is provision for the Fish and Wildlife service and the NMFS to pick out specific areas that are called protected critical habitat zones, which is aimed at reducing habitat loss seemed to be the primary threats of endangered or threatened species. There was however an amendment of the law in 1978 by Congress which made the designation of critical habitats mandatory necessary for all endangered species. With this amendment there was the inclusion of economic consideration in the process of picking out habitats for the endangered species which was to be run through on the basis of the available scientific data. There was later an amendment of the same law in 1978 which brought out conflicts due to the prevention of economic consideration by the 1978 amendment. Studies done between the year 1997 and 2003 stated that there was a huge possibility that endangered species in critical habitats we're twice as lucky and more likely to improve in population number and the recovery rates would only increase. This is due to the provision of all the essential needs of the species by the Fish and Wildlife service which has a policy in place that limits the designation of land and water within the United States Which states that all federal agencies can exclude the essential areas in the event that the costs of these essential areas exceed the benefit though the Endangered Species Act is quiet about the exact expenditure and benefits that are provided in determining critical habitats (Foreman, 2002).

All federal agencies are restricted from giving authorization on funding or going ahead with any actions that would compromise the condition of the critical habitat as stated in section 7A of the Endangered Species Act. Although the regulations of critical habitat are not directly applicable to private and non-federal landowners, Logging, and mining projects as well as not skill development being done on private and or state land need a federal permit and therefore, they become subject to the critical habitat regulations. The Fish and Wildlife Service and the NMFS with regards to the recovery plan have a mandate to formulate an Endangered Species Recovery Plan which outlines the purpose, likely costs, activities required and a projected timeline in the recovery of an endangered species. There is no specific information that talks about the time allocated to the completion of I recovery plan.

There is however a policy recorded with the Fish and Wildlife service covering this specific time allocation of a recovery plan which is to be within three years of the listing of the endangered species but the average time for completion of a recovery plan is estimated to be approximately 6 years. Recovery plans are critical aspect in the achievements of the ultimate goal of the law. With the amendment of the Endangered Species Act of 1988, Congress made an addition in Section 4 of the law that provided a clear line of the minimum contents of a recovery plan. A recovery plan has three types of information that must be covered in a recovery plan as listed below:

- Are clear description of site-specific management activities in the effort to bring out the plan as explicit as possible.
- The objective measurable criteria which serves as a basis for the judgment of when and how the recovery of the species is going on.
- A quotation for an estimated cost and the required resources in the achievement of the goal which is the recovery of a species and eventual delisting.

This amendment made an inclusion for public participation as an important part of the process. Similar to the listing process, there is a specific procedure and order that is followed for a recovery plan having the highest priority allocated to species that's are considered to be on the verge of extinction and would benefit most from the recovery plan particularly when the threat in question takes the form of construction, developmental or economic activities by man. Something to note is that recovery plans incorporates domestic and migratory species.

There are various exemptions that can be allowed under the Endangered Species Act only if there is consultation than with the United States Fish and Wildlife Service or the NMFS about any project that is to commence in a habitat of a listed species. This is trinitite a review by the agencies concerned in the event that the activity could impact a listed species in some way. The main dilemma to be sorted out is whether the listed species is in harm's way by the action and if it is how can the agencies minimize this harm. In the event the situation presents no possibility of avoiding harm to the species, the project agency seeks to proceed with an exemption from the endangered species committee which is composed over panel of members from an executive branch and also at least one appointee from the state which the project is to take place. There needs to be a vote of five out of seven committee members in order for the exemption plea to be allowed to take place which could include the pursuit, the hunting, the shooting, wounding or killing, harassment, or any significant habitat modifications towards the listed species (Foreman, 2002) (Figure 6.6).

Figure 6.6: A northern spotted owl.

Source: https://insaneowl.wordpress.com/page/58/.

6.2.5. Delisting Of Endangered Species Act

For a species to be delisted, there are several factors that are taken into consideration. Some of them are the controlling or elimination of the threat, the growth in size in population of the species, as well as the self-sustainability and stability of the habitat quality and quantity. There is also Special Situations and occurrences where over a dozen species have gone through the delisting process as a result of wrong data presented to the agencies having them being listed. There is also a term known as "down listing" of a species which means that the threats posed to the well-being of the species have been put under control and there was a recovery plan put in place that have successfully had the population meeting the recovery objectives therefore having the species moved to a different category from endangered to threatened. An example of an animal species that was recently delisted is the Virginia North flying squirrel which is a subspecies of the squirrel in August 2008 after it had been on the list of endangered species since 1985 (Nazzaro, 2010) (Figure 6.7).

Figure 6.7: A picture of the Virginia north flying squirrel.

Source: http://www.portlandmonthly.com/portmag/2011/06/no-fear-of-flying/flying-squirrel-encyclopedia-britannica/.

6.2.6. Effectiveness of Endangered Species Act

One would ask themselves how effective the Endangered Species Act really is. Studies have presented findings that 99% of all the species that are protected under the act have all dodged extinction. There is also proof of increase in population of these species and habitat development can also be registered making the results and effectiveness of the Endangered Species Act unquestionable. Due to the fact that it was passed as law in 1973 receiving bipartisan support, this categorizes the Endangered Species Act as one of the most effective laws when it comes to the protection of species that are facing extinction. As of January 2019, 85 species were delisted and a further 25 had been down listed from the classification status of being endangered to threatened. There has been argument that recovery of species that were threatened by Dichlorodiphenyltrichloroethane (DDT) for example the bald eagle, the brown pelican and the peregrine falcon, deserve to be accredited due to the ban on DDT by the Environmental Protection

Agency (EPA) instead of the Endangered Species Act. There is however an upside to the ban on DDT which instigated non-DDT oriented activities which is due to the efforts of the Endangered Species Act, such as habitat protection, disturbance protection as well as captive breeding of species (Nazzaro, 2010).

As of the year 2019 January records, there are 1467 species in total, both domestic and foreign on the endangered and threatened lists. There is however quite a number of species that have suffered extinction as they were candidates for listing, or they were under consideration for listing. Some of the species that have faced population increase in size under the Endangered Species Act are: The bald eagle, The Mexican Wolf, the Gray Wolf, the Gray whale, California's South sea otter, the whooping crane, the Big Bend *Gambusia*, the peregrine falcon, the Hawaiian goose, the grizzly bear, the Florida's Key bear, the Virginia big-eared bat, the Black-footed ferret, among many others (Sullins, 2001) (Figure 6.8).

Figure 6.8: A picture of the key deer.

Source: https://en.wikipedia.org/wiki/Fauna_of_Florida.

In Section 6 of the Endangered Species Act, there is provision for funding with regard to development of programs for the management of the endangered or threatened species under the state wildlife agencies. As a result, the list of endangered and or threatened species found within the states boundaries are prepared by each individual state. This is under the state endangered species list which includes different species that are considered endangered or threatened in a particular state, but they are not suffering the same consequences in other states. This therefore leads to them not being listed on the national list of endangered and or threatened species. Some examples are in states like Florida, Minnesota, and Maine.

6.2.7. Penalties

With this law, there are different levels and degrees of violations that can be committed by a person. The very seriously punishable violations are the trafficking and the knowingly taking with an inclusion of harming, killing, or wounding of an endangered species. Penalties that are imposed on these violations include a monetary fine of up to $50,000 or a jail term of up to one year or both. Civil penalties can have a placement of fines up to $25,000 per each violation. There is provision for the exact fines and the list of violations found on the National Oceanic and Atmospheric Administration website. There is an exemption placed on the violations if there is evidence that the violation was committed in self-defense therefore no charges are imposed. As an addition to the fines and the jail terms, there must be issuance of a permit or a license given by the federal agencies that grant permission for the importation of fish, plants, and wildlife can be revoked or canceled, modified or placed under suspension. A holder of a permit to hunt or fish, that has committed a violation of the Endangered Species Act stands a chance of having their permit canceled or placed under suspension (Sullins, 2001).

The use of the money gathered through these fines is presented as reward cash for the persons involved in the furnishing of information that leads to the arrest or the conviction of a criminal as well as the revocation of licenses provided that they are not a local, state or a federal personnel in the line of duty. Some of these funds are allocated to the care of fish and wildlife and forest service. In the event the balances exceed $500,000, the secretary of the Treasury is mandated to make deposits that is equal to the excess into the cooperative Endangered Species Conservation Fund (Sullins, 2001).

6.2.8. Challenges Faced By the Endangered Species Act

The successful implementation of this act has faced several challenges due to the constant misinterpretation of the Acts demands.

- Cost that are incurred be the industry. These bills present themselves in the form of lost opportunity or the slowing down of operations to agree and act in accordance with the regulations presented in the Endangered Species Act.

- The argument that the Endangered Species Act has a possible encouragement of preemptive habitat destruction or the taking of the listed species or proposed species for listing by private landowners. This was experienced at the time where one month in the scenario of his forest responded to the Endangered Species Act listing of the red cooked woodpecker and increased the harvest and short and the age on harvesting of his trees to ensure that they are not suitable for habitat for the species in question.

Due to these constant clashes, the Endangered Species Act is constantly put between a rock and a hard place of putting the interests of the conservation and protection of endangered or threatened species against the interests of the industry.

With regard to extinction and the risk of endangered species, it has been noted that despite the ultimate objectives of the Endangered Species Act, there is relaxation when it comes to species that are no longer listed which has happened repeatedly. In almost all of the 50 years of its existence, there is less than and approximated 50 species that have been delisted as a result of eventual recovery. There is no denying that since the enactment of the Endangered Species Act several species have since gone into extinction noting that they were listed under this act and their stones potential risk for species that are still listed and are protected under the Act. In the records dated up to January 2020, the essential services in this field indicate that 11 species have since gone into extinction and some of them are the Santa Barbara song sparrow, The Caribbean monk seal, the Tecopa pupfish, the blue pike, the Dusky seaside sparrow, the long jaw Cisco, the Guam broadbill and the Eastern puma. There is also a listing that has been done by the NMFS that consists of eight species that stand a very high risk of extinction in the predicted future. The biggest threat is human activity which poses a serious danger to the species themselves as well as the habitats that they live in. Their husband recorded 9 circumstances that have required the services to change the status of the species from straightened to endangered indicating that the species has moved closer to extinction. There has been critical evaluation and defenders of the act coming out to support and state that these claims are unfounded and hold no ground. This continues to be a serious fight do to the human activities consistently posing threats of the species and the non-compliance of individuals going against the Endangered Species Act.

Overall it can be claimed that there is a very huge success achieved annually due to the implementation and the enactment all the Endangered Species Act with the main objective of saving and protecting endangered species both in the United States of America as well as international nations where the act applies. Below are a few examples of recorded species that have had very high population increase do you to them being listed under the Endangered Species Act.

6.2.9. Success Stories of Species Under the Endangered Species Act

6.2.9.1 Black Footed Ferret

Figure 6.9: The Black-footed ferret.

Source: https://www.cmzoo.org/news/archive/the-keys-to-black-footed-ferret-breeding-success-lighting-testing-time/.

This is a 2-foot-long animal that comes from the weasel family and once existed in the central grasslands and basins of Southern Canada and Texas. In the early 90s it was seen to be one of the most endangered mammals in northern America. At the onset of the 1900s there was an estimated 5 million ferrets that were housed by the United States of America. The population decline was due to hunting by dogs for food which is considered to be species invasion and so this species reaches an almost wiping out in the early 20th century as the aggravated result about the cultural development and the introduction of rodent pesticides and poisons that completely devastated the whole population. A record three years after it being listed with the Endangered Species Act in 1967, there was the death of the last captive ferret which brought people to an eventual conclusion that the animal was extinct in North America. At the start of 1981, there was the discovery of a small population of this colony and it is recorded that between the year 1991 in 1999 there is an estimated 1200 ferrets that came as a result of the protection of the Endangered Species Act. Currently, it is estimated but there is now a total of 1410 ferrets in the wild (Figure 6.9).

6.2.9.2. The Red Wolf

Figure 6.10: Image of the red wolf.

Source: https://www.pinterest.com/pin/689050811710009595/.

This species was recorded and listed as endangered in 1967 with a population of only 17 red wolves in the wild. There was no development of a critical habitat and the recovery plan was completed in 1990 after its listing. They were breeded and released into the wild with positive results produced by their population which was a total of 300 by the year 2011 (Figure 6.10).

6.9.2.3. The American Alligator

Figure 6.11: The great American alligator in the wild.

Source: https://en.wikipedia.org/wiki/American_alligator.

The American alligator was listed in the year 1967 as endangered and the recovery plan was completed in 1979 from the date of its listing. Habitat loss and extremely who are regulations with regard to hunting was the primary factor that led to the decline of the species population. It later on recovered and was delisted in several states from 1975 all the way to 1987 (Figure 6.11).

6.9.2.4. The Gray Wolf

Figure 6.12: The gray wolf.

Source: https://www.nwf.org/Educational-Resources/Wildlife-Guide/Mammals/Gray-Wolf.

This species was listed in 1976 as endangered which was exacerbated by human activities of hunting and trapping eventually bringing down the population of the species. There was a collaboration between The American government and the Mexican government aimed at restoring and protecting this species between 1977 and 1980. There was a development of a federal recovery plan in 1982 and by the late 80s the recovery plan had been concluded. The total population stood at 309 Wolves by the year 2004 and in the year 2010 the population had increased to 333 Wolves (Figure 6.12).

6.9.2.5. The Short Nose Sturgeon

Figure 6.13: Picture of the short nose sturgeon under water.

Source: https://www.fisheries.noaa.gov/species/shortnose-sturgeon.

This species once existed in rivers and along the Atlantic seaboard nearly saw the face of extinction due to overfishing, by catch, river damming, poor water quality as well as habitat destruction by human activity. It was placed under the Endangered Species Act listed as endangered and has since increased in population by at least five times and to be more specific the population as it stood in 1979 when the species was being listed was 12,669 and today there is more than 60,000 of this species alive in the wild waters (Figure 6.13).

6.9.2.6. The California Least Tern

Figure 6.14: The California least tern.

Source: https://en.wikipedia.org/wiki/California_least_tern.

The decline in population of these birds came due to their hunting and shooting as a form of creating fashion decorative for women's hats in the early 19th century. The population also plummeted as a result of habitat destruction by the development and recreational pressure. It was listed as endangered in 1970 at a population of 225 in California. Due to the protection of the nests and beaches for development and disturbance their settlement and habitat, the population increased steadily and stood at 7100 in 2004 and in 2010 the population continued to increase that was recorded at 6568 species by the United States Fish and Wildlife service where it underwent the down listing process (Figure 6.14).

In conclusion, the introduction and the eventual enactment of the Endangered Species Act has done commendable work in the rescue, protection, and recovery of species that are considered to be endangered or threatened, and more often than not see the species through the entire process from its listing to its eventual delisting. The number of species that have been saved due to this act has been an outstanding amount and this continues to stamp the importance of environmental law with regard to every aspect that is considered as a composition of the environment. The countless amendments that created room for more species to be listed including plants is nothing but very impressive by this law. We continue to see improvement in the environment due to the application of this act in various situations and it has also created a relationship between the agency is concerned with implementing the Endangered Species Act and the people of the community.

This critical for this act to continue being implemented and carried out as law due to the fact that it helps the sustaining and maintenance of the ecosystem of which human beings, plants, and animal species I found it upon and depend on a daily basis. With the ecosystem being weak and unhealthy we continue to leave ourselves at risk and increase the vulnerability of natural disasters as well as the disasters caused by human activity to occur due to environmental degradation. Having this law stand strong is important for the medical world because majority of medicine is derived and originally developed from plants and natural products. It is also important for humans to acknowledge the fact that nature is beautiful, and it creates a form of recreational satisfaction in the lives of men on a daily basis. There's also an important role that is brought out by tourism when a nation or a region has strong wildlife and nature. This is important especially when there is economic gain in terms of revenue for nations that depend on tourism.

Agricultural sustainability is also very important as a branch of environmental law. With a poor environment the population of both human and animal species are vulnerable of hunger, drought, and starvation leaving no possibility for life to sustain itself in a certain region. This is key as it continues to instill and hammer down the importance of environmental law as an attachment to the Endangered Species Act. There is a relationship between the sustainability of the life of human beings and the life of the species. Our day-to-day lives are intertwined in a sense that, plants take him carbon dioxide that human beings exhale and they give out oxygen that is important for human being life support because they inhale oxygen. Birds are important we need to comes to the passing of pollen from one flower to another in a process called pollination which creates more and more life for plants and flowering plants to be specific which also extends life to insects like bees that take nectar from plants and create honey which is consumed by human beings. In this way we see that it is very important for all aspects of the ecosystem to remain alive and healthy in order for the neighboring species to sustain itself and remain healthy and strong.

There is also a key important part that is brought out by the act where species that are at risk of being extinct you to species invasion are protected. This evaluation and identification is important because it helps us regulate and keep under control the populations of animal species bird species and plant species in order to avoid usurping of one species by another. For example, if all the fish want to go extinct, the quality of marine water would decline substantially on the death of coral reefs which are considered to be very important in the ecosystem would suffer a severe blow. This also extends a monitoring of habitats of all species as it is important for the species to have a livable space in order for its population to avoid plummeting.

There is also a great need to encourage international conversations and conferences to embrace implementing the Endangered Species Act as law in different nations which is important for not only the United States of America and its peripheral nations and extensions, but for the whole world in general. Having the enactment of this act in many countries across the world only does more good and continues to restore and heal the environment as it is important for the very life that we hold dear as a human race.

6.3. THE CONVENTION ON INTERNATIONAL TRADE IN ENDANGERED SPECIES OF WILD FAUNA AND FLORA (WASHINGTON CONVENTION)

CITES (convention on international trade in endangered species of wild fauna and flora) is a global agreement between countries. Its purpose is ensuring that the international trade of specimens of wild fauna and flora does not affect their survival. Today, extensive information about the endangered status of various important species (such as tigers and elephants) may make the need for the convention obvious. But when the concept of citation was first formed in the 1960s, international discussions on the regulation of wildlife trade for conservancy purposes were relatively new. In hindsight, the need for citations was obvious.

Each year, the global wildlife trade is projected to be worth several billions of dollars, including hundreds of millions of animal and plant specimens. This trade is diverse, from live faunas and florae to various wild animal products gotten from them, including food, exotic leather products, wooden musical appliances, wood, travel antiques and medicine. The amount of exploitation of certain animal and plant types is rather high, while their trade as well as other factors (like habitat loss) may seriously consume their populations and even make some species endangered.

Many wildlife species used for trade aren't threatened, however the existence of agreements to ensure trade sustainability is important to protect these resources in the future. Since wildlife trade crosses borders between countries, international cooperation is needed to regulate wildlife trade to protect certain species from overexploitation. The convention on international trade in endangered species of wild fauna and flora was imagined in the essence of such cooperation. Today, whether traded as a living specimen, fur, or dried herbs, wildlife are given varying degrees of protection to more than 37,000 animals and plants.

The Convention on International Trade in Endangered Species of Wild Fauna and Flora was drafted in accordance with a resolution adopted by members of the IUCN (World Conservation Union) in 1963. The script of the convention was finally approved on March 3, 1973 at a representatives meeting of 80 nations in Washington, DC, and enforced on July 1, 1975. The originals of the Agreement were added in the depositary governments in English, French and Spanish, and each text has the same effect. The agreement is also accessible in Chinese and Russian.

For the reason of this convention, not unless the framework otherwise requires: (a) "species" here means any type of species, subspecies or geologically separated population; (b) "specimen" also means: (i) any fauna or flora, whether they are alive or not (ii) for animals: for species mentioned in appendixes (i) and (ii), any easily identifiable parts or derivatives; for species included in appendix iii, any products in appendix iii related to the species easy-to-identify parts or their derivatives; (iii) such as genus plants.

Appendix II of the Washington Convention covers (a) all species, although not in danger of extinction now, unless the trade of specimens of the species is strictly regulated to avoid incompatible use with survival; (b) The trade-in specimens of certain species mentioned to in paragraph (a) can be effectively controlled and must be regulated for other species. Appendix III includes any species that a Party has determined to be controlled within its area to prevent or restrict exploitation, including all species that need cooperation from other Parties in trade control.

7

Environmental Cleanup

CONTENTS

7.1. COMPREHENSIVE ENVIRONMENTAL RESPONSE, COMPENSATION AND LIBERTY ACT (CERCLA)

7.1.1. Introduction

This is part of the environmental laws that have been developed and regulated by the US Federal Government. It is also referred to as a superfund set up by the EPA. This law was established in order to set prohibitions as well as requirements when dealing with closed and abandoned hazardous waste sites. This was with an aim to generate a liability for individuals responsible for the release of the hazardous wastes in the given sites. This act was passed in 1980. This was as a result of the alarming deadly unacceptable hazardous waste practices and management that was being conducted in the 1970s. This is an environmental program that was set up in aims to assist in the management of hazardous wastes. The act was amended by the Superfund amendments and Re-Authorization Act in 17th October, 1986. This program ensures that the invaluable removal actions are taken. It also makes some enforcements on the parties responsible for the hazardous sites. It is in charge of creating the accountability, community involvement and the long term protectiveness of the environment from the hazardous wastes as well as contaminants endangering the environment (Bearden, 2012).

The main objective of CERCLA is to identify sites where the hazardous materials have been deposited and have a threat to the environment. The areas present a potential risk to the public health due to the leakage, spillage or any other mismanagement that may be conducted which includes the lack of good hazardous waste removal plans. After identifying the areas, they seek to find the party that is responsible for the dumping of the wastes so that they can take accountability for the wastes. After the identification of the responsible party, the next step involves the cleanup of the site. In most cases, the party responsible for the dumping of the wastes is the one conducting the cleanup process.

It is one of the ways in which they take responsibility for the wastes. The sites in reference are referred to as the Superfund sites. Though their parties responsible for the dumping of the wastes conduct the cleanup, CERCLA conducts authorization and supervision of the cleaning process. There are two major ways in which the authorization is conducted. They include the short term removal and the long-term environmental remediation. It is referred to as long term environmental remediation as it involves activities that seeks to protect the environment from hazardous wastes and also the healing of areas in the environment that have been affected by these wastes.

The actions conducted during the cleanup are done as per the EPA's National Priorities List. There are conditions that must be adhered to when undertaking these activities. The activities are to be conducted promptly to abate, prevent, reduce, stabilize, mitigate, and remove any potential threat to the environment and the public health. One of the conditions that is followed when dealing with these wastes is that the wastes being handles should be in the lists of hazardous materials and substances. They may also be any contaminants or pollutants and this is of course with the exemption of gas and oils. The body in charge of this environmental act is EPA. It houses the act and ensures that all the laws and recommendations are met. They are the group that hires and releases personnel to the environment to conduct evaluations thereby identifying areas that have been affected by the wastes or areas containing the wastes (Bearden, 2012).

The EPA makes the attempts to identify the parties responsible for the contamination. This step is important as it allows for the identification of the parties that are to take responsibility of the contamination. There are various categories of individuals responsible for the contamination. For instance, they may include the generators of the hazardous materials in the location. They can also be the transporters of the hazardous materials to the site or the past or present owners of the site. In most cases when dealing with the party responsible, they have to willingly undertake the responsibilities that have been task to them.

It is more like they have no option. This is because even if they decide not to comply with the tasks given to them EPA will still order them to do it. In a case scenario where they are unable to do the task any way or they refuse to comply with the EPA's orders then the EPA can initiate the response and removal actions by themselves. The funding for the actions is catered for by a trust fund which is made possible by the taxes that are imposed on the chemical and petroleum industries. This was done when CERCLA was being enacted. To some extent, one may state that the responsibilities were undertaken by CERCLA are similar to those of the Resource Conservation and Recovery Act, Safe Drinking Water Act (SDWA) and the Clean Water Act (CWA).

The undermining principle is that for the CERCLA, it focuses more on the hazardous wastes. It is more of its specialty. CERCLA and RCRA both share the same jurisdiction when it comes to matters management of hazardous wastes and the underground storage materials containing petroleum products. The protocols and guidelines followed when handling these wastes are under the Hazardous and Solid Waste Amendment. CERCLA is the body that to some greater extent manages these wastes because most of the wastes that are to be handled are outlined by it (Bearden, 2012).

7.1.2. Response Actions

As stated above there are two response actions that are undertaken by CERCLA. They include:

Removal actions: These are short term actions. These actions through the short term, are authorized by CERCLA. They are actions undertaken to address the releases or the threatened releases to the environment that may require drastic approaches or responses. They are said to be short term actions as they deal with the immediate responses undertaken once there is an identification of land containing hazardous wastes as well as contaminants.

Remedial actions: These are long term actions that are undertaken by CERCLA. They are also authorized by CERCLA. These actions are permanent and are said to significantly minimize the risk of the release of harmful substances. The condition for these actions to be undertaken is that they have to be under the national priorities list. For a site to qualify to enter the list it must first be subjected to a Preliminary assessment also referred to as a site inspection. It is afterward subjected to a Remedial investigation also referred to as a feasibility study.

Under the Act, apart from the two responses, there are also some liabilities that are underlined in the Act.

Liability: The parties responsible for the hazardous waste sites have are all in a wide net and brought together by CERCLA. In the scope of the liability act, CERCLA has the responsibility to facilitate enforcement. It states that," when the president establishes that there may be an imminent and substantial potential danger to the public health or the welfare of the environment due to the act the releases or threatened release of a substance from a facility." It has been said that there are changes being made to the act with regards to the retroactive nature. This is as a result of the responsible parties being held liable for the actions that predate the statute (Bearden, 2012).

In the courts, they have held on to the decision that if the problems resulting from the wastes continue and the statute made a reimbursement obligation the statute is not retroactive and therefore is constitutional. The court involved in this matter is known as the Monsanto court. It justifies the retroactive nature including CERCLA operating remedially to spread the cost of the response being made to the improper disposal of wastes. The cost is spread among all the parties involved; they played part in the creation of the hazardous condition. There some changes that were proposed under the commercial clause that have been rejected (Bearden, 2012).

7.1.3. Potentially Responsible Parties

There are parties that may have in one way or another be involved in the making of the hazardous condition. The identification of the responsible parties is done after the identification of the hazardous sites. In order to locate the potentially responsible parties, EPA looks for incriminating evidence so that they can determine the liability. This is done by matching wastes obtained from the sites with parties that may have contributed to wastes present in the sites. In most cases under the act, there are four entities that may qualify to be under the potentially responsible parties.

They include:

- The present owners. They may also happen to be the operator of the site. This may involve the excluding of property that may have been acquired through security interest. This implies that the current owner will still be involved though they may not have made any contribution to the hazardous release. For instance, in NY v. Shore it is outlined that any a landowner who have purchased a contaminated site within the intuition of cleaning and developing it will be held accountable for the cleanup cost as the owner was aware of the state of the land before he or she purchased it. This may include the fact that he or she was aware of the fact that dumping was still ongoing as the purchase was underway.
- The past owner. This involves the operator of the land in the time frame that pollution was said to be occurring. The initial owner of the land are also held accountable for the release of the toxic substances that occurred while the land was under their possession. The act states the party who does not or did not own a facility may be held responsible as an operator if they have no substantial control over the activities that contributed to the release of the hazardous substances.

- Individuals involved in the disposal of the hazardous wastes at a site. Under the useful product act, it is stated that if an individual is not arranging for disposal then he or she is out of the system. Therefore, the producer of a certain chemical that is sold for use is not an arranger.
- An individual who may have taken part in the transportation of the hazardous wastes to the site. A transporter liability is enforced on the transporter who in a way has played a role in the selection of the site where the wastes are taken and dumped. Also if the transporter takes part in moving the wastes from one of the contaminated areas of a parcel to another area which is uncontaminated then the transporter will be held responsible. This is because they have aided in the introduction of contaminants to an area that initially had no contaminants present in it. They have contributed to the spread of the hazardous condition.

When any of these individuals are found to be guilty in any of the listed forms, they are referred to as the potential responsible parties. The EPA once they have gathered enough evidence and identified these individuals, they are at liberty to involve them in any of the acts of responses. They are also among the parties that will be involved when incurring the costs due to the response actions that have been undertaken. In case where individuals feel that they are being forced into something that was not their responsibility then there are defenses that have been outlined to protect them.

7.1.3.1. Defenses

There are strict liabilities that have been laid out by CERCLA in conjunction with the Joint and Several Liabilities which have resulted in a major liability that is imposed in the relevant parties. For the parties that want to avoid the liabilities set out by CERCLA, they can use the available defenses though they are limited. Some of the defenses are outlined as follows:

The defendant may escape the liabilities if he or she is able to prove that though the individual had knowledge concerning the pollution, he or she took an initiative to reduce the probability of spills occurring. A good illustration is the shell case held in the Supreme Court. It was determined that Shell supplied chemical products to the Brown and Bryant Inc. was not an arranger in this case. Shell was, therefore, found not responsible for the spills in Brown and Bryant. The court also noted that there were steps taken by Shell to reduce the potential of a spill occurring. Also taken to accountability was the mere knowledge that Shell had in that spills and leaks contented to occur was not a strong basis for Shell to be held accountable as the arranger of the disposal.

There are other codes that also provide potential defenses. This may include the court's decision on the act of God also commonly referred to as an act of war. Most of the court have held on to the decision that some of the defenses do not apply in certain cases which involves events that could be anticipated. Some of these events include earthquakes, heavy rains, and hurricanes among others. The defenses are therefore limited to extremely rare and unforeseen circumstances.

There is also the third party defense which includes the defense applying in the case where the potentially responsible person is able to demonstrate that they exercised care needed by the environment. It also applies when there was no contractual relationship with the egregious third party. This implies that the defense will not be applicable in event that the dumper was an employee, agent, subsidiary, parent company, or anyone who was in contract with the PRP. For instance, the third party defense may apply in a case scenario of midnight dumper. In this case, there is an assumption that without the knowledge of the owner, someone goes and dumps the toxic waste secretly on the land of the landowner. The landowner, therefore, had no reason to suspect that any illegal dumping was being made on his land.

There is also a defense for the innocent landowners. This defense was as a result of criticism made on the law. This led to an amendment being made to CERCLA. The amendment is known as the Superfund Amendments and Reauthorization Act. The amendment catered for a defense being provided for the innocent landowners. The innocent landowners, in this case, refer to those who made inquiries on the land before purchasing it and they had no knowledge concerning the hazardous substance. The defendant in this case should be able to prove all these facts.

There is a defense for the De minims settlements. In the case where the EPA establishes information that there was a PRP contribution to a minimum amount of hazardous waste in comparison to other wastes present in the site and the waste did not contribute effectively to the cleanup then the EPA may be forced to offer settlement to the PRP. The policy was set aside more so for business owners who may have not been major contributors to the hazardous contributions. In the case of contagious properties there is a code that protects the neighbors to the polluted areas. In the case where there is an individual who takes up responsible care and is unaware of hazardous wastes leaking in the individual's land from a contagious property, the owner of the land may not be held responsible.

`When dealing with lands having the hazardous waste there is a name given to it to allow for proper identification. Under the act, the name given to this area is Brown fields.

7.1.3.2. Brown Fields

A brown field is the name given the property that may be complicated due to the presence or potential presence of a hazardous substance, pollutant or contaminant. This area is not fit for human beings to dwell in as it poses potential threat to their health as there are multiple ways in which the contamination may slip in to the human body. This potential of land having harmful substances on it has made developers and other investors refuse to buy land. This is because they fear the cost it may incur purchasing and cleaning the land. Amongst their fears may be the potential liability that may be imposed on them as a result of owning land.

This resulted to the congress creating a code in order to protect the developers and potential land owners. By so doing, it encourages them to engage in activities that aim at cleaning the brown fields. There is a statue in place that prevents liberty being assigned to a bona fide prospective purchaser. This is done on the condition that the potential land buyer will not impede the performance of either o response action or national resource restoration.

Initially, the cleanup was accounted for by the taxes from the consumers of petroleum and chemical products. Later on in a congress, the cost was shifted and the burden laid on the taxpayers who happen to be the citizens of the country. In recent years, there have been reports that the program was suffering from underfunding. This explains why most of the Superfund activities have reduced significantly over the years. In a bid to calculate the score of the site the EPA as well as the state agencies make use of the Hazard Ranking System to do the calculation. All the calculations are done on the basis of the actual or the potential release of hazardous substances from a site.

In the case where a site gets a score of 28.5 then it qualifies to enter the list of the National Priorities List. It also qualifies for a long term remedial action within the Superfund program. The number of sites being identified keeps on increasing. The act also authorizes the federal natural resources agencies the major one being the Environmental Protection Agency (EPA), states, and Native American tribes being able to recover from natural resource damages that are as a result of hazardous substances. The different states have different versions of the Superfund law.

Under CERCLA, there is the Agency for Toxic Substances and Disease Registry. There are a variety of activities that are conducted which involve cleanups, placing of engineered controls and also the creation of site restrictions including groundwater use restrictions.

The second goal may be to restore the site to its initial state whereby it can be productive and can be used for a variety of activities including the land being converted into business areas, recreational parks or natural ecosystems. If one is able to quickly identify the intended, use of the land then the process will be much faster and the cleanup process will be less expensive. The added advantage is that the Superfund Redevelopment Initiative caters for the provision of tools as well as the necessary support required of the site redevelopment (Banks, 2018).

Due to the liability scheme created by CERCLA, there have been changes that have impacted the commercial and industrial real estate. This is because they become more liable for contamination even from past activities. This implies that they cannot pass liability to unknowing buyers without any responsibility. The buyers also have to be aware of future liabilities. There was need for the revision to be done on the National Oil and Hazardous Substances Pollution Contingency Plan. This program deals mostly with the response to the release and threatened releases of hazardous substances, pollutants or contaminants. They are the group that established the National Priorities list.

This list serves as EPA's information and management tool. The list is usually updated by the federal rulemaking. One of the major roles conducted by the EPA is the identification of the hazardous waste sites (Bearden, 2012).

7.1.4. Use of the NPL in Site Identification

The identification of a site for the NPL is purposely intended to guide the EPA in:

Determination of sites that warrant further investigation for the purpose of assessing the nature and extent of the risk to the environment as well as the human health. They also identify the CERCLA-financed remedial actions that are appropriate. They notify the public on the areas (sites) the EPA believes that it may require further investigation. Therefore, the public is able to avoid these areas and if they were to purchase land they would first wait for the investigation to be conducted before and further contracts or agreements are made (Rostron, 2013).

They also notify the PRPs that the EPA may begin a CERCLA-financed remedial action. Because the PRPs may be involved in the financing of these projects, they need to be notified first so that they can prepare themselves financially. They are also involved in the cleanup process so the notification of an upcoming projects is necessary for them.

The inclusion of a site in the NPL does not necessarily mean that the PRPs have to initiate the cleanup actions nor assign the liability to any person. It is the duty of the NPL to serve information purposes thereby notifying the government as well as the public of the sites and also the releases that may seem to be warrant to the remedial actions.

There is a difference between the authority to address hazardous substances and pollutants or contaminants is the cleaning up of the pollutants and contaminants which may not be hazardous substances and cannot be compelled by the unilateral administrative order. Though CERCLA is referred to as the Superfund, there have been problems of lack of sufficient funds to clean up a number of sites in the NPL. This has led to the EPA negotiating the consent orders with PRPs so that they can study the sites and come up with cleaning alternatives. It has also led to the subjection of the EPA oversight and the approval for the involved activities (Rostron, 2013).

The EPA gives an issue on the Proposed Plans for the remedial actions for a given site which may need public comment and afterwards it makes a cleanup decision in a Record of Decision. The RODs are implemented under consent decrees by PRPs. If the consent cannot be reached, then there is the utilization of the unilateral orders. This involves a party being fined in a scenario where they refuse to comply with the order. If there is a party that spends money to clean up a site, they have the freedom to sue other PRPs in a contribution action under CERCLA. The CERCLA liability has been established in the judiciary as joint and several among PRPs to the governments for the cleanup costs.

The CERCLA liability can be allocated among the PRPs in contribution on the basis of the comparative fault. In case of an orphan share this is a case scenario where by the share of costs at a Superfund site is attributed to a PRP which is either unidentifiable of insolvent. The EPA has the aim of trying to treat the PRPs fairly and equitably. In light of budgetary cuts and constraints may make more fair treatment of PRPs relatively difficult (Lazarus, 2008).

7.1.5. Procedure Followed By the Superfund

When there is a notification of a potentially hazardous waste site, the EPA does a Preliminary Assessment which involves the records of reviews, interviews, visual inspection and also the limited field sampling. The information obtained from the PA/SI is utilized by the EPA to develop a Hazard Ranking System score. The score is utilized by CERCLA to know the status of the site. If the site gets a relatively high score, it is qualified to enter the list. The site may proceed to a Remedial Investigation commonly referred to as a Feasibility study. The Remedial Investigation involves an extensive sampling program as well as a risk assessment which will define the nature and the extent of the site contamination and the risks (Rostron, 2013).

The FS is utilized in the development and evaluation of various remediation alternatives. The most suitable alternative is presented in a Proposed Plan. This is done so that the public gets a chance to review it and also give their thoughts concerning it. After this is done there is the selection of an alternative in a ROD. Afterward, the site enters the Remedial Design phase than later on the Remedial Action phase. Most of the sites may enter the long-term monitoring. On completion of the Remedial Action phase the reviews are required as often as every five years. This is done when hazardous substances are left onsite above the levels that are safe for unrestricted use (Rostron, 2013).

There are a variety of tools that are utilized during the process of restoring the land to its initial stage. They include the following:

- **The CERCLA Information System:** This involves a database that is managed by EPA and the relevant states. It contains a list of sites where there are releases of the hazardous wastes have occurred. The areas that have been addressed and the areas that should be addressed. There are three inventories in CERCLIS. They include the CERCLIS removal inventory, CERCLIS remedial inventory and the CERCLIS enforcement inventory.
- **The Superfund Innovative Technology Evaluation:** This is a program that supports the development of technologies that will be used in the assessing and treatment of wastes in the superfund sites. The EPA conducts an evaluation on the technology and afterwards gives an assessment of the potential of the technology being utilized in the future Superfund Remedial actions. SITE

program is made of four related components. They include the Demonstration program, the upcoming technologies program, the monitoring and measurement technologies program and the Technology transfer activities.

- **The Reportable Quantity:** This is the minimum quantity of hazardous substances when released, they should be reported.
- **A Source Control Action:** This is a representation or installation and a startup of actions required for the prevention of continued release of hazardous substances to the environment. The wastes may be from the top or within the ground or in buildings among other structures. There is also the section 104e letter which is a request by the government for information concerning a given site. The letter may include a general notice to the PRPs informing them that CERCLA-related activities may be undertaken in a given site and the recipient may be held responsible. In this section there is the authorization of the EPA to enter the facilities and also obtain information on the PRPs, hazardous substance release and the liability and also an order access for the CERCLA activities. This letter is said to resemble written interrogatives in a civil litigation.
- **There is also the Section 106 Order:** This is a unilateral administrative order that is issued by the EPA to PRPs for them to perform remedial actions in a Superfund site. This is done when the EPA establishes that there may be an imminent and substantial potential danger to the public health or the state of the environment. This is very important as there is an actual potential threat posed to the environment if the given orders are not obeyed.
- **A remedial Response:** This is a long-term action which may minimize and stops the release of a hazardous substance which may affect the environment or the public health. The term remediation can at times be replaced with the terms remedial action, removal action, response action, corrective action or remedy. Under it there is the nonbinding allocation of responsibility. The term nonbinding implies that the person does not necessarily have to comply with the given responsibility. The NBAR is a device that was established within the superfund Amendments and Reauthorization Act. This gives the EPA the opportunity to make a nonbinding estimate of the proportional share that the individual responsible parties in a Superfund site pays towards the cleanup.

The relevant and suitable requirements are those that the United States Federal or state cleanup requirements may not be applicable in addressing problems efficiently similar to the ones encountered in the CERCLA site where their use may be relevant as it will significantly help in the cleanup of the area.

The Requirements in the given site may be relevant and appropriate if they would be applicable. This is with exception of the jurisdictional restrictions related to the requirements.

7.1.5.1. Implementation

It has been known that initially, over 70% of the Superfund cleanups were funded by the PRPs. In the case where the PRP is not available or is unable to fund the cleanup then the Superfund law caters for the cleanup through the taxes that have been imposed on chemical and petroleum industries. The taxes collected were intended to be utilized in the provision of incentives in the use of less toxic substances. The taxes collected go to the trust fund and are utilized in the cleaning up of abandoned and uncontrolled hazardous waste sites. Most of the money that has been collected ends up being used in the cleaning up of these sites as most of the PRPs have not been able to pay for the cleaning exercise. The cost has been catered for by the general fund appropriated by the congress.

7.1.5.2. Hazard Ranking System

This is a system that has been utilized in the evaluation of a site through the use of scores. The scores are able to identify whether a site has relative risks to public health and the environment due to releases or threatened releases of the hazardous wastes in the given sites. In the Superfund program, the EPA and state agencies make use of the HRS in the calculation of the scores on the basis of the actual releases or the potential releases of wastes in the given areas. The release may be through air, groundwater or surface water. As stated above, any area that gets a score above 28.5 gets qualified to be placed in the National Priorities List.

7.1.6. Environmental Discrimination

There have been federal actions that have been set up so as to address the disintegrating health of the environment and health. There are disparities concerning the given issues. The minority and low income populations are challenged with the disparities due to the Executive order that requires federal agencies to make environmental justice at the center of their policies and programs. The superfund sites have affected greatly the minor communities. Though there is a legislation that has been designed specifically to ensure there is equity in the Superfund listing, the marginalized population still experiences smaller chances of being listed in the NPL. This is comparison to the higher earning income areas (Beder, 2013).

When the Executive order was put in place there was a persistent discrepancy present between the demographics of the areas who lived near the toxic waste and their listing in the NPL as Superfund sites. This would have granted them an opportunity in the chance of being funded in the cleanup projects. The communities with the low income and the minority in the population were found to have reduced chances of site listing after the executive order this is in comparison to the higher income areas that were found to have higher chances of being listed in the NPL implying that their areas having the toxic wastes had higher chances of being cleaned up in comparisons to the areas of the minority (Banks, 2018).

There has been a study that has proven the presence of a slow response by the government to the community demands form the minority part of the community compared to the white communities. The Superfund sites found near the white communities have been said to have better chances of good cleanups, harsher penalties for the polluters and relatively larger tax-based funding compared to the minority. There are some case studies that can prove this theory. They include:

7.1.6.1. Case Studies in African American Communities

In the 1970s there was a case of the rural black community in Tirana who were found to be contaminated with PCB and DDT. Some of them had the highest records of these contaminants. For instance, for DDT the highest record was found in Indian creek which was relied upon by several residents for fishing. Though this area poses a major threat to the public health and environment, the contaminants were discovered in 1978 but the government acted on the issue five years later.

Another case may be for the West Dallas where most of the communities that lived there were African Americans and Latinos. The city officials were informed that children were being exposed to lead contamination. In 1974 the lead smelters were sued by the City and two years later the lead regulations were reduced. For a case study to be conducted, the residents had to wait till 1981 where the EPA commissioned a study to be conducted. The results that were obtained are similar to the ones that were obtained earlier. There was presence of lead that was exposing families to contamination. In 1983 there were cares that were shut down due to the lead exposure that was made possible by the smelters that were still operational.

Further in the investigation, it was discovered that the EPA deputy administrator had stalled the cleanup of the lead contaminated hotspots. In 1993 that the area was declared to be a Superfund site. It was among the largest sites in the list. In the Afton community of Warren County was among the major environmental cases in history. It is often referred to as the roots of the environmental justice movement. In this area, PCB's were being illegally dumped into the community and as a result the area became a PCB landfill. The community leaders mounted pressure on the state concerning the cleanup of the site for a decade when their hard work bore fruits. The site was cleaned up. Despite the cleanup, the site did not regain its initial state. This has led to the community calling for reparations to be made.

The last case concerns the Bay view-Hunters which is a historical African American community. This area has encountered persistent environmental discrimination which is as a result of the poor remediation efforts of the San Francisco Naval Shipyard. There was negligence from various agencies with regards to cleaning up of the site. This meant that most of the residents were prone to pollution and gentrification. This may be elaborated health wise by the high rates of cancer and asthma among other diseases that have been recorded by most of the residents. The area requires to be included under the NPL or it would imply that most of the individuals will have to suffer in terms of health as a result of the present pollution.

7.1.6.2. Case Studies in Native American Communities

A good illustration is of the Church Rock Uranium mill spill that occurred in Navajo nation. By far this is the largest radioactive spill that has been encountered in the US. Although it is the largest, the response by the government has been relatively slow as the site is referred to as a low priority site. The US Congress has put in place the cleanup plans. Though these actions have been undertaken, the areas have not been significantly cleaned. To date, the uranium contamination from mining continues to pose a health risk to the residents in the area.

If cuts were to be made to the EPA funding, it would mean that the cleaning would be stagnated for long periods of time. It would, therefore, hinder the regulation and remediation of the Superfund sites. The delays being made by the government in response to the cleaning of these areas would mean that the residents of these areas will continue to be exposed to health risks due to the present contamination. If drastic measures are not taken by the government then several of its citizens will be sick and much of the government money will go into treating them. It Al's means that the economies of such countries will be affected as there will be no revenue coming in as the citizens are not working.

The EPA funding is very important because in cases where the government response to the cleaning of these areas is slow, the EPA will be able to step in and make relevant steps towards ensuring that the areas are being rejuvenated to their original state.

7.1.6.3. Assessing of Data

The data utilized and present in the superfund is usually available for the public. This implies that the public is able to access data concerning the Superfund program. Some of the data available to the public include the EPA Superfund information systems that entails reports and product descriptions. Under it there is information concerning the Superfund Product Order Form. There is the TOXMAP which is a geographical information system from the division of specialized information services. It was deprecated on 16th December, 2019 in the United States National Library of Medicine. The application utilized maps of the United States to aid in the visual exploration of users as they deal with the relevant data from the EPA Toxic Releases Inventory and Super fund programs. The TOXMAP is a resource that was funded by the federal government. TOXMAP's chemical and environmental health information is obtained from NLM's Toxicology Data Network and PubMed among other sources.

When dealing with CERCLA, one has to look into various areas which are associated with what it does. This includes the Brownfield land which is discussed above and is among the areas taken under consideration in the NPL. Other areas that may be taken under the NPL is the formerly used defense sites. This are areas that have been used by the US state's soldiers.

7.1.6.4. Formerly Used Defense Sites

This includes properties that were initially owned or possessed by the United States and other jurisdictions owned by the United States Secretary of Defense. It also referred to as the U.S military program that was created in 1986. It was established with the aim of conducting assessment as well as environmental restoration that may have been led by the U.S Army Corps of Engineers. One may look at it as a way in which the army cleans up its mess. There is a lot of money that has been put into military training, production, weapon installation and its testing. There are large tracks of land that is under the U.S military. With regards to CERCLA, there are over 9800 sites that have been identified that need cleaning in the case of military training sites. The sites have been identified under the Department of Defense. Most of the identified sites were said to be in urgent need of cleanup. There are restoration projects that have been underway and some of them have already been initiated (Bearden, 2012).

When it comes to these areas, there are some regulations that have been outlined by the Defense Environmental Protection Program statute. They work in conjunction with CERCLA in conducting direct assessment, the eligibility for cleanups and cleanup. It also does the National Oil and Hazardous Substances Pollution Contingency Plan. For the site to qualify for a funding, the condition in place is that the contamination should have occurred prior to 17th October 1986. It is still the duty of the EPA to determine whether the site qualifies to be in the Superfund site listed in the NPL (Bearden, 2012).

In a bid to get more understanding concerning the CERCLA and dealing with the transporters of the wastes in the sites.

7.1.6.5. Hazardous Materials Transportation Act

One of the ways in which an individual can be considered as a PRP is if they are in one way or another involved in the transportation of the waste from a facility to the identified site or transporting the wastes from one area of the site to another area of the given site. When dealing with the transportation of hazardous wastes there is the following of the guidelines outlined in the Hazardous Materials Transportation Act. The act was enacted in 1975 and is the principal federal law in the U.S which regulates the transportation of Hazardous materials. It was laid out to ensure the protection against the risk to life, property, and the environment that are inherent in the transportation of hazardous material in intrastate, interstate, and foreign commerce (Bearden, 2012).

The act was passed so that it can facilitate the improvement of the uniformity of the existing regulations involved in the transportation of hazardous materials and also aid in the prevention of spills and illegal dumping which would lead to the endangering of the public and the environment. This is a problem that has been faced by most uncoordinated and fragmented regulations. The act involves the regulations being enforced through the use of four key provisions that encompass the federal standards. The four areas of key provisions include the procedures and policies, material designations and labeling, packaging requirements and the operational rules. If an individual chooses to violate the HTMA regulations, they may qualify to be given a civil or criminal penalty in the exception of a case scenario where there was a permit that was not granted under the discretion of the secretary of Transportation (Sands, 2003).

As stated above there are two types of penalties that may be given to an individual who violates the HTMA. One of the penalties include the civil penalties. Under this penalty, an individual is required to pay a certain amount of money once they violate the regulation, order, special permit or approval of the Act that have been done knowingly. There is special violations that are taken into consideration for each day that the violation is done by the individual who is involved in the transportation of the hazardous material and the transportation goes on. The individual is said to be transporting the materials knowingly if:

The individual has knowledge on the facts related to the violation or if the individual acting in the circumstances demonstrates reasonable logic and care would have that knowledge.

Before the civil penalty is given to an individual, there are a variety of things that the secretary takes into consideration. They include the nature, circumstance, extent, and the gravity of the violation. They also take into consideration the degree of culpability of the violator. They check if the individual has a history of violations, their ability to pay and the potential effect on the ability to continue doing the business. The violator is given a chance of having a hearing. They are also entitled to a notice given to them by the secretary specifying the amount of penalty (Sands, 2003).

In the criminal penalty, there is also a certain amount of money that the violator is charged with. This is in the case where the individual knowingly tampers with labels or packages used during the transportation of hazardous material. This involves an individual who willingly or recklessly violates an order, regulation, special permit or approval outlined in the act being fined, getting imprisonment of not more than five years or both. The violation may lead to the release of hazardous material that may cause body injury or death of an individual. This may lead to an individual getting a sentence of ten years.

The tampering, in this case, may occur in different ways. It may involve the alteration, removal, destruction, or unlawful tampering of the following: a marking, placard, label, or description on a document that has been prescribed by the regulations of the Act. It can also be the tampering of a package, component of a package, packaging, container, rail freight, aircraft, motor vehicle, or any material that may have been used in the transportation of the hazardous material. An individual is said to act knowingly if the individual has knowledge on the facts related to the violation or the person acting in the circumstance demonstrates reasonable logic and care would have the knowledge.

An individual is said to act recklessly if they display a conscious disregard to the consequences of the individual's conduct (Atapattu, 2007).

7.1.7. Special Permits and Exclusions

Like other acts, there are special permits and exclusions that are offered to the individual. The secretary may give off an issue that will modify and terminate a special permit authorizing a variation to regulations. The special permits were initially effective for not more than two years. The renewal of the special permits may be granted if the secretary gives a discretion once there is an application that has been field for successive periods of not more than four years each.

There are applications of the special permit. For one to apply for the special permit the individual should be able to provide a safety analysis that has been prescribed by the Secretary who then does the justification. The applicant then submits the application. The application is made to the Administrator of the pipeline and hazardous material Safety Administration. Afterwards the secretary does the publishing of a notice of the application to give the opportunity for public review. The secretary gives an issue, renew or deny the application within a period of 180 days (Bearden, 2012).

In the case where the secretary will delay the filing date, he will give a publish statement to the federal Register addressing the reason for the delay. They also give the possibility when the special permit will be given off. Once the secretary finishes the review of the circumstance of the given permit. They give the public an opportunity for the comment and review. They give a new rule that allows the incorporation of the special permit in the regulations of the Act. There are various sections in which the special permits are applicable. The following is a description of the relevant sections where the sections are applicable. The regulations for safe transportation are for individuals who: transport hazardous material in commerce, causes to transport hazardous material in commerce, those who design, inspect, repair, test packages that have qualified for hazardous material transport, individuals who prepares or accepts hazardous material for transportation in commerce, an individual who is responsible for the safety of transporting hazardous materials in commerce or an individual who certifies a compliance with the requirements written in the act (Bearden, 2012).

There is also the representation and tampering regulations for a package, component of a package or packaging that was intended to be utilized during the transportation of hazardous material. The shipping papers and disclosure regulations and the highway routing of hazardous material regulations.

When one discusses about CERCLA, one needs to have knowledge on the National Oil and Hazardous Substances Pollution Contingency plan as there is a correlation that exists between it and CERCLA. They both deal with hazardous wastes (Bearden, 2012).

7.1.8. National Oil and Hazardous Substances Contingency Plan

Also known as the NCP. It is part of the United States federal governments plan when responding to oil spills and hazardous substances releases. In this document there is the national response capability which was set up with an aim of promoting overall coordination among the hierarchy of responders and contingency plans. The first National Contingency Plan was created and published in 1968. This was done as a form of response to a massive oil spills from an oil tanker off the coast of England. The oil spill involved over 37 million gallons of crude oil being spilled in water. It resulted to major environmental damage.

In a bid to avoid problems encountered during response by the officials in the incident, the US officials came up with coordinated approaches so that they can cope with potential spills in its waters. In 1968 the plan facilitated the first comprehensive system of accident reporting, spill containment and its cleanup. There was the establishing of a response headquarters, national reaction team and also regional reaction teams. This was followed by the broadening of the scope of the National Contingency Plan through the years. NCP was revised the following year to include a framework that deals with the response to hazardous substance spills and oil discharges.

With the passing of the Superfund legislation, the NCP was broadened so that it could cover the releases at hazardous waste sites that will require emergency removal actions. There are other additions that have to the NCP to allow them keep up with the enactment legislation. The recent revisions to the NCP were done to reflect the oil spill provisions of the Oil Pollution Act. In the National contingency plan, there should be plans made by the federal agencies for emergencies and the developing of procedures that will address oil discharges and the releases of hazardous substances, pollutants or contaminants. This will allow them coordinate their planning, preparedness, and response activities with the affected states (Bosselmann, 2016).

A response is triggered and the USCG is given an authorization to initiate and in cases where there has been a discharge that poses significant threat to the public health or the welfare of the United States is needed to initiate and direct responses that appropriate activities as the Administrator determines any oil or CWA hazardous substances is discharged. In the case where there is a substantial threat of a discharge from a given vessel or offshore or onshore facility into waters, it may affect the natural resources that belong to the United States. It is the duty of the federal on-scene coordinator to direct response efforts and coordinates all the efforts at the scene of a discharge or release (Bosselmann, 2016).

There are significant revisions that were made to the plan. They include in 1968 the Initial plan, added provisions for hazardous substances in 1973, revision to reflect CERCLA provision in 1980, revision, and reorganization to reflect SARA provision in 1990 and finally the revision to reflect OPA revision in 1994. In case of an oil spill there is a significant role played by BP. The plan involves responsibility being placed on it so that it can allow for the command as well as the control when managing serious disaster responses with the US federal government and not a private company like BP. There are some major provisions that are involved when dealing with these plans. For instance, there is the establishing of the National Response Team and the roles and responsibilities that it plays in the National Response System.

This includes the planning as well as the coordination of responses to the major discharges that have been made for oils or hazardous wastes. They also provide guidance to the response teams. They coordinate a national program of preparedness planning and response. They also facilitate research so as to improve the response activities. In this case, the EPA serves as the lead agency in the National Response Team. In section 300.115, there is the establishing of the Regional Response Teams and the role they play. They have a responsibility in the National Response System which involves the coordination of preparedness, planning, and also response while in the regional level.

The RRT is made of a standing team made up of representatives of the federal agencies and are members of the Net, state, and local government representatives. The RRT plays the role of providing oversight and consistency review for the area plans in a given region. There is the establishing of the general responsibilities of the federal On-Scene Coordinators in section 300.120. It is needed that a notification of any discharge or release to the National Response Centre through a toll-free telephone number given to the public in case the cases occur. The National Response Centre plays the role of the central clearing house for all pollution incident reporting.

There is also the authorization of the predesigned On-Scene Coordination that is direct in all federal, state, and private response activities at the areas of discharge. There was the establishment of a unified command structure for the managing responses to discharges or threatened discharges through coordinated personnel as well as the resources of the federal government and the responsible party. In section 300.165, it is required that the On-Scene Coordinator submits to the RRT a report concerning all the removals that were taken from the site (Bosselmann, 2016).

7.1.9. Limitations Faced in the Amount of Funds Used for Specific Activities

Every program that has been set is said to encounter some challenges. The challenges cause a setback in the progress being made by the project in place. With CERCLA, there are challenges that have been encountered some of which have led to some set goals not being achieved. The biggest problem of all deals with the funding of the sites. Funds are necessary as they facilitate most of the activities conducted by CERCLA. The biggest challenge has been lack of funds. Some of the states decided to reduce the amount of money that was being allocated to the Superfund. This being the case, it has forced the Superfund to look into other ways in which they can generate money to fund the cleanup. This has been through the use of PRPs. Though the use of PRPs has been quite significant, the major challenge has been that most of them have been unable to comply with the orders given to them as they lack the necessary funds that is required. This has led to the Superfund using money that was collected as tax from chemical and petroleum industries (Bearden, 2012).

In the recent day, they have been given money that has been collected from taxpayers. Though this step has been taken, most of the sites listed in the NPL have not been cleaned due to lack of funds. The number of sites keep on increasing year after year meaning that the Superfund becomes more limited when using their funds. The Superfund has therefore resulted to giving more priority to the areas that have the highest scores. These are areas that are highly likely to pose health problems and the destruction to the environment due to the wastes that are present in the area.

Apart from funds, the other challenge that may have been encountered with the Superfund is the slow response by its officials and the government. Though they are notified of the presence of hazardous wastes in a given site, they are taking too long to conduct a study or even clean up the site. This means that the risk of residents being contaminated keeps on increasing. In some cases, even though the problem has been identified, some of the officials are stalling the cleanup process in some of the sites. This delays the cleaning in most of the sites (Bearden, 2012).

8

Waste Management

CONTENTS

8.1. RESOURCE CONSERVATION AND RECOVERY ACT

8.1.1. Introduction

The resource conservation and recovery act was created in 1976. It is an act that governs solid, hazardous, and municipal waste disposal by federal law in the United States. In full it is an act that makes provisions for assistance that is technical and financial to help in plan development for management and establishments for energy recovery and other resources from materials that have been disposed of and for the proper disposal of the said materials safely and also regulate hazardous waste management. The law was enacted by the 94th Congress of the United States and it was effected on the 21st of October 1976. It is an amendment to the solid waste disposal act of 1965. The act was brought forward to the senate by Jennings Randolph in 1975. It was considered by the public works committee of the senate and the committee of house interstate and foreign commerce and it was passed by the senate on the 30th of June of 1976. It was signed to law in 1976 by President Gerald Ford on the 21st of October in the year 1976.

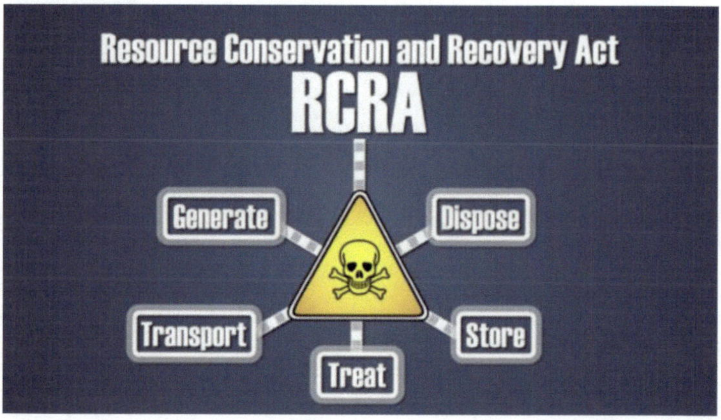

Figure 8.1: Common aspects or areas of the resource conservation and recovery Act.

Source https://www.convergencetraining.com/blog/what-is-rcra.

In order for the environmental protection agency (EPA) to reach their goals from the resource recovery and protection act, they have a reliance on three main programs that is the regulatory framework that has been established, partnerships that are collaborative with multiple stakeholders and the program for corrective action (Figure 8.1).

The RCRA has various components that is the act itself which gives a detailed description of program that are related to managing waste that are to be established by the congress and authority is also given to the administrator who has office in the EPA. It also gives mechanisms for the establishment of regulations or standards as authorized by the act which are as promulgated by the EPA. These regulations are then published federal register. The EPA has also developed guidelines for procedures to be applied in the process of implementing the act and finally policies in relation to the Act.

8.1.1.1 History and Goals

The resource conservation and recovery act was created to solve and address the problems faced by the country in relation to the increasingly growing volume of solid municipal waste and together with that being produced from industries and as a result it was an amendment from the 1965 act on solid waste disposal.

8.1.1.2. Goals of the Resource and Recovery

- The protection of the health of humans and the environment from hazards that have the likelihood of causing harm;
- Conservation of energy and resources present in the natural environment;
- Waste reduction in relation to the amounts generated and also through recycling waste; and
- Ensuring waste is managed in a manner that is environmentally sound.

The resource conservation and recovery act is a joint partnership between the federal state and the EPA of the United States and providing state requirements to adapt, adopt, and enforce the Act. It is one of the most widely known regulations for the management of waste and was developed under a set of standards for hazardous waste treatment, storage, and disposal. The act also plays a major role in managing municipal and industrial waste and also the management of underground storage tanks.

8.1.1.3. Implementation

The EPA has published regulations on managing waste which are coded in title 40 of the code of federal regulations at parts 239 through 282. Regulations that are related to the management of waste that is hazardous began in part 260. Based on act states have been given the authority to develop and manage their own programs on hazardous waste and are similarly required to be as strict as the ones by standards from the federal government. The different states have also been given the task of implementing plans for solid waste management.

8.1.2. Provisions of the Resource and Recovery Act

8.1.2.1. Subtitle A – General Provisions

Objectives of the resource and recovery act: The objectives of the act were mainly to offer protection of health and the environment and also conserve material and energy resources by:

- Provision of assistance either technical or financial to state, local governments and interstate agency to foster the development of plans for solid waste management to promote improved techniques for the management of solid waste, improved, and new methods for solid waste collection, separation, and recovery and the safe disposal of those that cannot be recovered in a way that is environmentally sound.
- Provision of grants for training in professions involved in the design, operation, and maintenance of systems for solid waste disposal.
- To prohibit open dumping of waste of land in the future and also requiring the conversion of dumping sites actively in use into waste management facilities especially those that have the potential to cause danger to the environment and the health of humans.
- Regulation of the hazardous waste treatment, storage, transportation, and disposal especially those that can cause adverse effects to the health of humans and cause harm to the environment.
- Promotions of programs for national research and development to improve the management of solid waste and development

of modern techniques for the conservation of resources and making organizational arrangements that are more effective and methods of new improved methods for the processes involved in the management of waste and the safe disposal of practices and systems of residues that are non-recoverable.

- Promoting solid waste management through demonstration, construction, and application ad resource conservation of techniques and technologies which preserve and enhance air, water, and land quality resources.
- Establishing efforts cooperatively among the federal government, state, local authorities and private enterprises in order to recover materials that are valuable and the energy from solid waste.

Government cooperation: The interstate cooperation-the provisions by this subsection of the act that are to be carried out by states can also be carried out by agencies that are interstate, this means that activities carried out by states can also be carried out by interstate regions where such agencies and regions have been established by the respective states and are approved by the administrator (The United States Environmental Protection Agency, 2014).

Consent of Congress to compacts-it is given to two or more states to negotiate and enter into agreements or compacts in order not to conflict with any law in the United States for mutual assistance and an effort that is cooperative for solid and hazardous waste management or both and at the same time enforcing laws that are related and also establishing agencies and joints as they see fit for making agreements and compacts effective. No agreements or compacts shall be obligatory or binding upon any state, a party thereto unless there is an agreement among parties interested and it should also be approved by the administrator (The United States Environmental Protection Agency, 2014).

Application of the Act and Integration With Other Acts: According to the act there shall be nothing in the act that shall apply to any activity or substance that is related to the water pollution act of the federal government, the act for safe drinking water, protection of marine life, act for research and sanctuaries of 1972, act for atomic energy in 1954 except in the event where the application is not inconsistent with the requirement of the above acts.

Integration of the resource and conservation act with other acts-provisions of the act integrated for the purpose of administration and enforcement to avoid its purposes from being duplicated up to a point that is practical with appropriate provisions to other acts. The integration of acts is expected to be effected only to the extent that it can be done in a manner that is consistent with the expressed policies and goals in the act and other acts that are referred to in the acts that are mentioned in the subsections of the Act.

Solid Waste Management Information and Guidelines: This subsection provides guidelines for the management of solid waste which are to be adopted in consultation with the administrator, agencies of the federal government, the state, municipal, and inter-municipal agencies and other parties that are interested and after public hearings have carried out to develop and publish the guidelines that have been suggested (Figure 8.2).

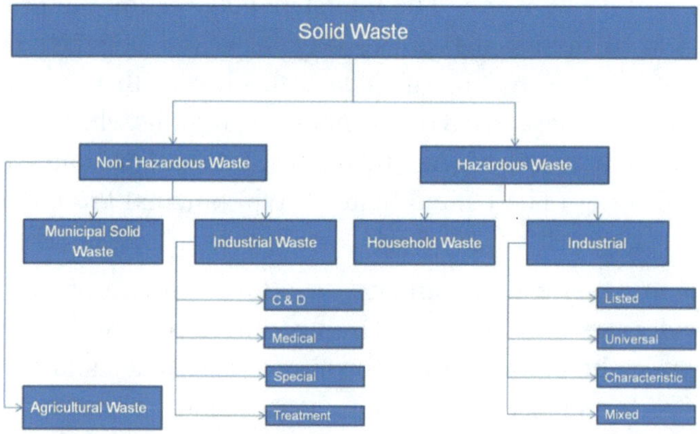

Figure 8.2: Classification of solid waste.

Source: https://en.wikipedia.org/wiki/Resource_Conservation_and_Recovery_Act#/media/File:Solid_Waste_Types.jpg.

The suggested guidelines shall:

- Provide information and description of the level of performance that can be attained by practices for solid waste management through giving a technical and economic description to provide protection to the environment and human health.
- It also describes the levels of performance to be expected after enactment including methods that are appropriate for control and monitoring methods that provide at a minimum the protection of the public and its welfare, water quality protection for both

surface and groundwater, air quality protection especially that which is ambient, the control of diseases and vectors and finally ensure safety and preserve the aesthetic value of an environment.

- It also provides criteria to be used as a minimum to be adopted by states in defining practices for solid waste management which includes dumping of wastes in open areas that is either solid or hazardous waste which are to be prohibited by the fourth title of the Act.

8.1.2.2. Subtitle B – Office of Solid Waste; Authorities of the Administrator

The administrator in the case is expected to create an office at the EPA for solid waste which is to be headed by a deputy administrator of the EPA. Therefore, the duties that were expected to be carried out basing on this act would be carried out through the same office.

Authorities of the administrator: The administrators have been tasked with different functions or has the authority to do the following:

- The administrator has the authority to prescribe or make suggestions on regulations that are necessary for him or her to carry out functions that is in consultation with relevant authorities such as federal, state, and authorities found regionally.
- Make consultations with or make an exchange with regards to information with other federal agencies taking part in research, development, and demonstration of projects or studies that are related to the management of solid waste and also their programs.
- Provide assistance that can either be technical or financial to agencies at either state or regional level while developing and implementing management plans for solid and hazardous waste and also programs that are of relevance.
- The administrator will also be expected to make consultations with representatives from different organizations such as research and education institutes, industrial sector, the labor sector and organizations that deal with protection of the environment and similarly consumers.
- The administrator is also expected to properly make good use of information, facilities, personnel of other resources of federal agencies including the states bureau of standards and the bureau

in charge of census, on a basis that it is subject to compensation in order to perform research and analysis and similarly conduct studies and investigations related to the recovery of resources and their conservation and also carry out responsibilities that are related to the act or under the Act.

Regulations that are made under the act will be reviewed or be revised if necessary and therefore can be done frequently which is not less than three years.

Resource Recovery and Conservation Panels: This section of the act gives the administrator the power to create teams or personnel with inclusion of different and relevant authorities and can be referred to as panels for resource recovery and conservation to provide assistance that is technical if requested by the government on the management of solid waste, the recovery of resources and also their conservation. The teams shall include different professionals including technical, marketing, financial, and institutional specialists and the said services shall be provided without expecting payment from the state or the government agencies that requested the services.

Grants for Discarded Tire Disposal: The administrator is also required to make grants available which is equivalent to 5% of the purchase price of tire shredders to those applicants who have been considered using a criteria for eligibility. Applicants eligible for the grant include private purchasers, public bodies or joint ventures that include both public and private agencies. The criteria for receiving grants shall be directed by the Act. For public bodies, the administrator is required to determine the qualification and ensure that no private purchasers are not interested in making the application before approving the grant to the public body. The administrator was also required to grant up to 750,000 dollars in grants every fiscal year to implement the section.

Annual Report: For each fiscal year, the administrator is required to submit a report or transmit a report to congress and the President which is to be given not later than 90 days after the fiscal year. The act requires that the report submitted gives information that is comprehensive and also detailed on the activities that are carried out in the office during the year that comes before it. Some of the aspects that are required to be included in the report are a statement with details on the objectives for the activities and programs that were carried out under this act, statements making conclusions by the administrator as to how effective the program and activities in meeting the objectives that were stated and the purpose of the Act. The report should also give a summary of the problems related to the management of solid waste that were met while the administrator was working and this should be in order of their priority. The report should also contain future plans the administrator has for the following fiscal year.

General Authorization: provisions from this act require the appropriation of funds to the administrator based on the act and are to be determined by the government every fiscal year for the purposes of carrying out duties that are stated under this Act. The act also states that not less that 20% of the amount allocated shall be used for panels formed under the act and their expense shall be incurred by funds allocated based on the Act. According to the act not less than 30% of the amount allocated for the purposes under the act shall be used in relation to the management of hazardous wastes which are stated under the Act.

8.1.2.3. Subtitle C – Cradle to the Grave Requirements for Hazardous Waste

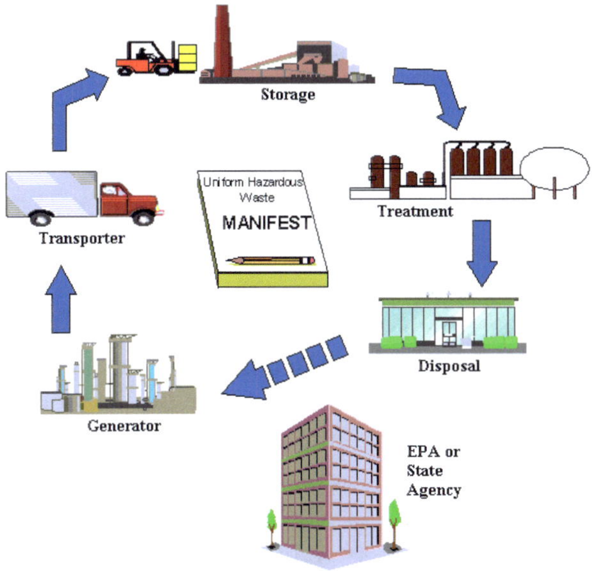

Figure 8.3: Cradle to grave system for management of waste that is hazardous.

Source: www.savannahenvironmental.com/cradle2grave.htm.

Provision for hazardous wastes: This part of the act makes requirements for managing wastes that are hazardous. The first section describes criteria for listing wastes that are considered as hazardous and this should not be done later than one and a half months of the act being enacted which is done together and in consultation with parties previously mention in the Act. The criteria are based on the toxicity, persistence, and the nature of degradability and its potential for accumulating. Other factors include the corrosiveness or the flammability of the waste (Figure 8.3).

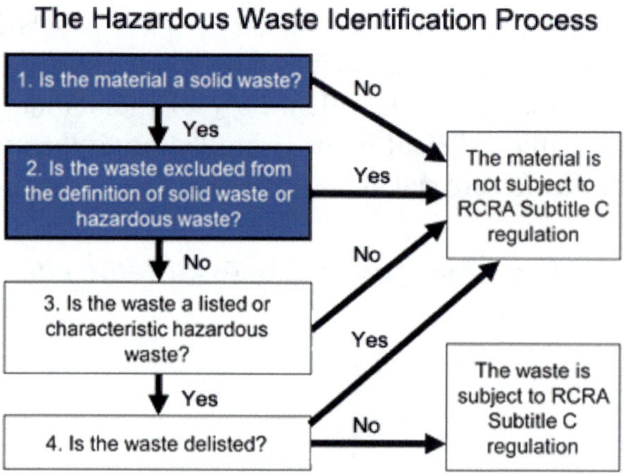

Figure 8.4: The criteria for determining and listing waste as hazardous.

Source: https://www.epa.gov/hw/criteria-definition-solid-waste-and-solid-and-hazardous-waste-exclusions.

Hazardous wastes are classified using lists that is F, K, P, and U lists. Wastes that fall in the F list result from industrial process in manufacturing, K list wastes are industry specific while the P and U lists contain waste from unused chemicals that are either pure or commercial (Feinberg and Foley, 2009) (Figure 8.4).

For the purpose of identifying waste or classification, they are assigned codes by the EPA. For the F, K, P, and U list they are assigned either of the four codes that are characteristic of waste that are considered hazardous.

- Toxic Waste (T);
- Acute Hazardous Waste (H);
- Ignitable Waste (I);
- Corrosive Waste (C);

- Reactive Waste (R); and
- Toxicity Characteristic Waste (E).

Parameter	Defining Characteristics	Example
Corrosivity	Waste which have pH < 2.0 or > 12.0 or which corrodes steel at a rate greater than 6.35 mm per year at 55°C.	Acidic waste, spent pickle liquor.
Reactivity	Wastes, which are unstable and spontaneously react with water or air, generate toxic gases and explode due to shock or heat.	Water from TNT operation and used Cyanide solvents.
Ignitability	Wastes which spontaneously ignite in dry or moist air at or below 60°C.	Waste oils, used solvents.
Toxicity	Wastes which release toxic materials on leaching in sufficient amounts to pose a substantial hazard to human health or environment as measured by the Toxicity Characteristics Leaching Procedure (TCLP)	Metals bearing wastes.

Figure 8.5: A table classifying the characteristics of hazardous waste and their description.

https://www.environmentalpollution.in/waste-management/hazardous-waste-definition-and-classification-industries/2798.

After assessing the waste to determine whether the waste is hazardous, they are listed if found to be hazardous under the provisions made by this act which will be reviewed or revised time after time whenever it is found to be necessary. A petition can also be made by the state governor 18 months after the enactment of the act to identify a list of materials as hazardous. After receiving the petitions the administrator is required to act upon them after a 90-day period after which he or she is required to notify the governor and also provide a statement to the governor on the considerations made by the administrator (Feinberg and Foley, 2009) (Figure 8.5).

The act also makes provisions for standards applicable to the generation of wastes considered to be hazardous. After 18 months of enacting the administrator is also required to develop standards with respect to keeping of records which identify the amount of waste generated, labeling practices, appropriate containers for management of hazardous waste, chemical composition of wastes that are hazardous, appropriate designs for facilities that manage hazardous wastes, report submission to the administrator if an agency has the permit to manage a facility dealing with hazardous waste.

It also gives provisions for standards applicable to hazardous waste transporters for example recording keeping for hazardous waste transportation, labeling of waste that is hazardous, compliance to manifest systems, transportation of hazardous waste to waste that deals with their treatment and permit requirements. Regulations on transport of hazardous waste shall be coordinated with together or in consultation with the secretary for transport. The regulations are also required to be consistent with the requirements other the Act. The administrator is also required to make recommendations the transport secretary respecting regulations that are under the act for the transportation of wastes that are considered to be hazardous (The United States Environmental Protection Agency, 2014).

The act also gives provisions for standards that are applicable to owners and operators of hazardous waste that includes waste treatment, storage, and disposal in facilities. This includes the maintenance of records, compliance monitoring, and reporting, operating methods and techniques, the design and construction of the facility, plans for contingency in case of accidents and damage, maintenance operations and also other compliance requirements. Such facilities will also be required to show financial responsibility or consistency in their operations.

The act also gives provisions for permits with regards to facilities for treatment, storage, and disposal of hazardous waste this includes requirements for the application of permits, provisions for the issuing of permits and also the revocation of permits as a result of not complying to the stated regulations and also the interim status for those who have applied for permits. The act also gives guidelines for developing authorized programs for the management of hazardous waste by states or the federal government (The United States Environmental Protection Agency, 2014).

Inspections: Inspection of waste management facilities are to be done by a member of the EPA as designated by the administrator or upon the request of any state employee. Those authorized to carry out such inspections have the authority to check for compliance e.g., for checking labeling requirements, checking plans or testing samples and this is to be done in a considerable amount of time. Records from inspections should also be made available for access by the public upon being certified by the administrator.

Federal Enforcement: The administrator is also in charge of compliance orders for example giving violation notices to facilities that do not comply with regulations for example by the state. If a compliance violation is issued by the administrator followed by an order for suspension or license of revocation, the persons involved are given a grace period of 30 days before a decision to revoke their license is final and that is the right to a public hearing. The compliance orders issued by the administrator should also bring forward the violation that is of concern. If the violations are done willingly the persons involved shall be convicted, penalized or be fined for each violation or imprisonment that does not exceed a year or both depending on the judgment made (Banks, 2018).

8.1.2.4. Subtitle D – Non-Hazardous Solid Wastes

The objectives of the subtitle are meant to be of assistance in the development of methods for solid waste disposal that are friendly and sound in relation to the environment, at the same time make use of resources that are valuable and also foster the conservation of natural resources. The said objects are similarly to be accomplished by the provision of either technical or financial assistance. The guidelines are also established to create cooperation among involved and interested parties.

Provisions for Waste Management Plans: Guidelines for waste management plans are provided by the act this includes the guidelines for regional identification for the purposes of planning which is done after 80 days of enactment. It entails identifying areas that have similar waste problems and the guidelines depend on the location and area size, the amount of generated waste and the means available for coordination of activities. It also provides guidelines for developing state plans and also provides considerations for the same depending on the characteristics of wastes.

The act also gives provisions for minimum requirements for the approval of plans for the management of waste that is non-hazardous this includes the responsibilities in the implementation of state plans, prohibition of the establishment of new sites for disposal, closing or the upgrading of disposal sites, establishment of powers for regulation by the state and plans for the conservation of resources that are valuable. Part 4004 of the subsection contains the criteria for landfills that are sanitary and this includes the developing a criterion for the landfills, the type of waste required to be in landfills. The subsection also gives provisions for the upgrading of waste disposal sites that are open that is by providing regulations for the inventory in waste management sites or the closure and upgrading of dumping sites that are existent. This are activities required to be performed by the state as directed by the administrator in order to prevent or minimize the potential of waste to cause health hazards (The United States Environmental Protection Agency, 2014).

The administrator has the power to approve or disapprove state plans 6 months after they have been submitted this if the set requirements are met, revisions can be made as the administrator sees fit in order to bring the plans into compliance. It also gives provisions for determining whether a state or agency qualifies for financial assistance if they meet requirements within a given period. Under this subtitle the administrator can approve applications by states. If plans for a state are withdrawn by the administrator the technical and financial (Banks, 2018).

Under this subtitle the federal government is required or expected to provide funds to assist in carrying out activities that are under the act and afterwards the administrator is also expected to provide financial assistance to states, municipalities or related agencies. The applicants for financial assistance for funds is expected to comply with the plan that is agreed plan. The administrator is responsible for allocating funds to states for every fiscal year and will depend on the population ratio of the state. For the purpose of developing or implementing state plans the administrator maybe required to provide technical assistance with respect to the resource recovery and conservation act using panels that are created by the EPA. The administrator also has the responsibility of identifying communities in respect to solid waste management for examples those with environmental problems that are serious. The administrator is also to be allocated money that is to be used for grants (The United States Environmental Protection Agency, 2014).

Rural community assistance shall also be offered to communities with populations of less than 5000 people or counties with greater than 10,000 people and those that are not within a metropolitan area. The administrator is also expected to allot funds appropriately every fiscal year and this should be in accordance with the regulations that are made under the act and should be based on the population ratio with that of rural areas for each state. The amounts allocated for each grant should not be in excess of 75% of the project costs, no assistance shall be provided for acquiring land for the use of setting up facilities dealing with the management of waste.

8.1.2.5. Subtitle E – Duties of the Secretary for Commerce That Is Related to Resource and Recovery

Functions of the secretary: According to this section, the secretary for commerce is expected to encourage the use of technology for the commercialization of the recovery of resources that has been proven. This can be done by providing specifications that are accurate for materials that are recovered, stimulating market development for materials that are to be recovered, promoting the use of technology that has been proven and finally the creation of forum to facilitate the exchange of data that is either economic or technical in relation to facilities for the recovery of resources (The United States Environmental Protection Agency, 2014).

After a public hearing the secretary of commerce shall be expected to publish guidelines for the development of classification of materials that are recovered from waste. This is done based on the collaboration with other state or government agencies like the national bureau of standards and other national organizations that set standards. The specifications are made based on the chemical or physical properties of the material of the waste. The collaboration will continue with organizations that set objectives and this will encourage the publication of these standards and also making updates for use of materials that are recovered (Banks, 2018).

Two years post-enactment of the act the secretary may take necessary actions such as geographically locating existing markets that have the potential for the commercialization of materials recovered from waste, identification of barriers that can either be economic or technical in the use of materials that have been recovered and encouraging the new development in uses of materials that have been recovered. According to this section of the act, the secretary of commerce also has the authority to make evaluations in relation to the feasibility of facilities that deal with the recovery of resources and thereafter publish the results from the evaluations that were made in the process. The results are then used in the development of a database that is purposefully used to provide assistance for those who select a given system for recovering resources (Feinberg and Foley, 2009).

8.1.2.6. Subtitle F – Federal Responsibilities

Guidelines for Applications of Federal, State or Local Law by Federal Facilities: Every department or agency or instruments of the arms of the government that have jurisdiction over the management of solid waste or are involved in any related activities to the disposal of wastes that are hazardous or solid are subject to compliance with the set requirements at regional, state or federal level that includes procedural or substantive requirements, with respect to control and abatement of the disposal of hazardous or solid waste. They will also be subject payments or service charges that are reasonable. Exemption of any facility that deals with managing waste from complying with requirements will be determined by the office of the president if it is in their best interest to do so. The exemption shall also not exceed a period of one year but extension of the period of exemption from compliance can or may be granted by the president still not exceeding a one-year period. At the begin of the year the president is expected to present to congress all the exemptions that have been made with respect to this section of the act and also provide a reason for giving the exemptions (Banks, 2018).

Procurement by the Federal Government: Procurement agencies are expected to comply with the requirements that are made under this subsection of the act and the item should not exceed the set amount per fiscal year. Any procurement done that is subject to regulation made by the administrator will not be expected to comply with this subsection of the Act. Two years post-enactment of this act, those involved in procuring materials shall be expected to purchase items that compose of the largest amount of materials that can be recovered practically. And in a manner that is consistent and to a level of completion that is satisfactory (Banks, 2018).

The decision made to procure the said items is based on whether the items are available within a period that is reasonable, whether the items to be procured fail to meet the set standards for performance and if their availability is at a reasonable price. Facilities that use fossil fuels to generate energy or for mechanical purposes are also encouraged to use materials recovered as fuel to the extent that is practical or applicable. Procurement specifications are also subject to review by federal agencies and the reviews functions to assess whether items comply or violate the specifications that have been set and this shall be done within a period of one and a half years after adoption and implementation of the Act.

The administrator after period of times will also revise guidelines to be used by procuring agencies which is to be done in consultation with the administrator of services that are general, the secretary of commerce via the bureau of standards and the printer for the public. Procurement services by agencies are also expected to be done in a way that will maximize the recovery of energy or resources. Policies addressed in this subsection are to be implemented by the administrator in collaboration with the procurement policy office that exists within the executive office of the president. It shall be the responsibility of this office within the executive to integrate procurement policies with other policies in a manner that leads to the maximization of the use of resources that are recovered from processes in waste management. Report will be presented annually to congress based on the actions that are undertaken by agencies at federal level which are required to show the policy implementation progress (Sancin, 2012).

Cooperation with the EPA: As the administrator carries out his/her functions permitted by the act, federal agencies that deal with waste management are expected to cooperate with the administrator to a full extent in order to fulfill the goals or objectives that have been set by the Act. Resources shall also be made available to the administrator upon request and this include personnel for instance technical personnel from the EPA, information or even facilities on a basis that is temporary.

The Applicability of Guidelines for the Disposal of Wastes That Are Solid to Agencies That Are Executive: An executive agency is any agency that has jurisdiction over a facility that manages activities related to waste that has been collected. If an agency is involved in waste management then the facility will be required to comply with guidelines that are recommended under the act for administration of the property or facility. If waste management activities are conducted by another person other than the executive agency, then they will be required to apply for a permit or a license. If an executive agency makes use of federal property for waste disposal use, then they will also be required to comply with the set guidelines for the purpose of conducting activities that are provided under this Act. Regulations that are to be carried out this subsection are to be prescribed by the president. In the issuing of permits or licenses, the executive agencies given the responsibility or those that bear the authority to do so shall consult with the secretary to ensure compliance before issuing licenses and permits for solid waste disposal. The guidelines for compliance are found under section 1008 of the Act.

8.1.2.7. Subtitle G – Miscellaneous Provisions

Employee Protection: This subsection of the act provides guidelines to protect employee rights which includes the right not to be discriminated against, prevent employees from being fired without valid cause especially if the employee or representative files a case that is instituted under this act or under an implementation plan that is applicable. Similarly, this applies to any individual that testifies or is expected to testify in any proceeding that is as a result from the implementation, enforcement or administration of this Act.

If an employee or employee representative believes that they have been fired unfairly or discriminated against by an individual perceived to be violating of part of the subsection mentioned above then, he or she has the right to apply for a review by the labor secretary 30 days post the occurrence of the violation who is expected to investigate the matter. A public hearing shall also be permitted upon request and based on the findings a decision will then be made and action taken as the secretary for labor finds appropriate. The labor secretary can either reinstate the employee if a there is a violation and if there is no violation then the secretary will deny the application of the employee or the employee representative (Sancin, 2012).

If the secretary of labor finds that a violation has been made, then the secretary of labor will determine the costs that were incurred by the applicant including the application will be assessed against the violation committed. There shall be no application permitted for employees who act against the directive of his/ her employer especially if they intentionally go against any requirement that exists in the Act. The administrator by provisions made by this act are expected to evaluate whether adopting requirements made by this act will affect the employment status of individuals. If an employee is set to loss employment as a result of the act or be discriminated against as a result of implementation of the act, then they will be entitled to a hearing after requested for an investigation by the administrator. After investigations are made into the issue then the administrator will make appropriate recommendations in relation to the matter. The report on the issue whether discrimination or layoff will be made a public record. The provisions by the subsection do not give authority to the administrator to make any changes with regards to plans for implementation of the act, standards or limitations (Macrory, 2008).

Citizen Suits: Any individual or group of people have the right to sue an individual or the government or any other agencies related to the government who is in violation of any requirement that is made by this act after it has been effected. They can also go ahead with civil action against the administrator in the event he/she fails to perform duties as required or expected by the Act. Any civil suit that is based on this subsection shall or will be brought forward to the district court in the district of violation of the Act. The jurisdiction shall within the district court regardless of the nationality of those in violation of the Act. Action to be taken against an individual, the administrator or state with regards to violating requirements, standards, permits, orders are not required to begin until the end of a 60-day period after a notice for violation is served or sent to the plaintiff. Those in contempt of the provisions made by the act are also given the right to intervene upon the commencement of proceedings in relation to the violation (Macrory, 2008).

The administrator is the one required to commence such proceedings after the 60-day period after the plaintiff has been served. The procedure for giving notice for action against a plaintiff shall be determined by the administrator. The administrator may also intervene in respect to the violation if he or she does not have direct involvement. The court also has the power and authority to award the reimbursement of cost incurred during the process of litigation by parties pursuing the lawsuit as seen appropriate. These fees may include the amount of money used for acquiring the services of a lawyer or an expert witness. The rights of those involved will not be restricted based on this section but it will be line with any common law that has the role of enforcing a standard or requirement related to solid waste management or wastes that are hazardous or go after relief (Macrory, 2008).

Imminent Hazard: If there exists imminent danger or substantial endangerment of human health supported by evidence as a result of activities related to handling, storage, treatment, transportation or disposal then, in this case, the administrator may bring suit on behalf of the country or state in the court district that is appropriate. This is done immediately to prevent the involved parties from continuing to take part in activities related to waste management that can cause harm to the environment or the health of humans if it is found to be necessary. A notice shall be provided by the administrator in the event a state is going to be affected by a lawsuit. Guidelines for the participation of the public in such processes will be determined by the administrator in cooperation or collaboration with the state involved. If circumstances for lawsuits are made invalid based on the act, the decision will not affect other similar suits (The United States Environmental Protection Agency, 2014).

Judicial Review: Any review by the judiciary of final regulations promulgated pursuant to the Act shall be in accordance with sections 701-706 of title 5 of the united states except that:

- A petition to review actions of the administrator provided under the act shall only be done in the District of Columbia in the court of appeal for the United States. After promulgation of a review, a petition can only be filed then after a 90-day period. The actions of the administrator will not be subject to review or criminal enforcement with respect to which review would have been obtained.
- In a judicial proceeding if an interested party files for leave to adduce additional evidence and shows that it is satisfactory to be submitted to as evidence then the court can order presentation of the evidence as found fit by the administrator. He or she can also change his findings and his recommendation if necessary by

setting aside the original order after new evidence is brought to light.

Grants or Contracts for Training Projects: The administrator has the authority to make grants to and contracts with organizations that are eligible. Eligible organizations refer to agencies that are interstate, state, municipalities, educational institutions and other institutions with the capacity to carry out projects effectively which can be funded with the grants provided under this subsection of the Act. The purpose may of this subsection may be fulfilled with use of grants to cater for partial or full costs which will be determined by the administrator. The cost catered for include projects operated by organizations that are eligible which are designed for program development, expansion, and performance of a program. The program may combine training, education, and employment of individuals involved managing, design, supervision, operation, and the maintenance of the disposal of wastes that is solid, facilities, and equipment used in the process of resource recovery. It can also be for the purpose of training instructors or those in positions as supervisors to improve their capabilities in supervising or training personnel involved in disposing solid waste and recovering resources through activities such as design, operation, and maintenance. No application will be approved unless the grant provides similar procedure and report required by section 207 of the act (The United States Environmental Protection Agency, 2014).

The administrator will also make investigations to determine the need for personnel to require additional training at the state or locally to carry out plans that are stated under the act and similarly programs for the management of solid waste, recovery of resources or both. Use of programs that are existent to train working personnel and finally determine if obstacles to employment and advancement in occupation exist, their nature in the field of waste management and recovery or resources which as a result may lead to limitations in man power that is available or personnel advancement in the said field. The administrator will then provide a report from the results of the investigation and study and this includes the recommendations of the administrator which are presented to both congress and the president.

Payment of grants can be made in advance or through reimbursement and in such installments and in such conditions the procedure will be determined by the administrator. There are no grants that will be made to organization that are private and whose main aim is profit-making as stated by this subsection of the act (Feinberg and Foley, 2009).

Labor Standards: Under the act, no grant for a construction project will be made unless the Secretary determines the application is supported with assurance that is reasonable that all employees are covered by the act of Davis-Bacon will receive wages that are similar to those working locally as determined by the labor secretary in accordance with the act and this will be in respect to labor standards.

8.1.2.8. Subtitle H – Research, Development, Demonstration, and Information

Research, demonstration, training, and other activities: The administrator will conduct and encourage, cooperate with and offer appropriate assistance financial or technical to public authorities, agencies, institutions, and individuals in the promotion of research, investigations, experiments, training, demonstration, surveys, public education programs and studies related to:

- Any adverse effects that may result from the release of materials that exist in solid waste to the environment and methods that can eliminate the effects;
- Solid waste programs for disposal operation and financing;
- Planning, implementation, and operations of systems for waste management or resource recovery and marketing of products that have been recovered;
- Production of forms of recovered resources that are useful;
- Developing methods of waste management that are new and improved;
- Identifying materials that can be recovered from waste;
- Developing technology for waste management and resource recovery that is small scale and low;
- Developing methods for performance improvement; and
- To make improvements in practices in land disposal in relation to the management of solid waste.

The administrator will also establish a management program for those pursuing the act to insure the coordination of activities related to research and development and also hasten the process of developing technology that is sound from the phase of carrying out research, through development and to the phase of demonstration. The administrator will also see to the development of pilot facilities to be used in testing the feasibility of technologies.

The administrator may also make grants or enter into contracts or contracts with authorities, agencies or private persons when carrying out part of this subsection that is in relation to carrying out studies, research, development, and demonstration. Contracts related to research and development will be made in accordance to research contracts of departments in the military. If any invention is created in the course of carrying out research under the contract shall be subject to the act of non-nuclear energy research and development and administered by the energy research and development administration shall be a substitute to the EPA. In this case solid waste will be substituted with the word energy. The personnel to serve this purpose will be determined by the administrator (Feinberg and Foley, 2009).

Special Studies, Plans for Research, Development, and Demonstration: This subsection of the act the administrator will publish studies and reports that determine recovery of resources from glass or plastic waste, determine the composition of solid waste streams, determine priority areas in study the process of resource recovery, small scale and low technology studies, research on the compatibility of front end source separation systems, study of the adverse effects of wastes from mining, conduct studies and publish reports on sludge. The administrator will also facilitate research on tires from discarded motor vehicles, effective function of facilities involved in recovering resources and the administrator will serve as a chairperson in the committee for resource conservation.

Coordination, Collection, and Dissemination of Information: The administrator will take part in developing, evaluating, and coordinating information in the management of solid waste that is the methods, practices, the amount of resources that have the potential of been recovered, existing technologies, managing hazardous waste, financing methods, market availability for resources that have been recovered and finally information on research and development.

The administrator is also required by the act to develop a library for reference on information collected. The administrator will also be required to develop a model accounting system that can applied to state and local governments in carrying out functions that are related to the management of solid waste. The administrator also has the authority to develop model codes for management of solid waste and also information programs to facilitate information dissemination on management of solid and hazardous waste and resource recovery. The administrator is also required to cooperate with state and local authorities in the process of collecting and passing on information.

Full Scale Demonstration Facilities: The administrator can also enter into agreements with agencies or private individuals to facilitate construction and operation of facilities for demonstration at full scale. After ten years of the act being enacted, the administrator will not be obligated to offer grants for the development of full scale demonstration facilities. The cost for developing full scale facilities can be shared among federal or state agencies as determined by the administrator in a manner that is practical or also desirable. After the enactment of this section of the act, the administrator will not be allowed to construct any facilities that are full scale in exception if it is by contract.

Special Study and Demonstration Projects on Recovery of Useful Energy and Materials: For action that is legislative or administrative, the administrator will conduct studies and similarly develop recommendations in resource recovery or either materials or energy on actions to reduce generated waste, methods used in processes of waste management, using incentives that are recommended, effect of public policies that exist and the effect of modification of incentives or disincentives related to the use of recycling of materials. Other recommendations include the necessity and methods of imposing charges in processes related to waste management and whether it reflects on the disposal costs and components of items that are recoverable, barriers to acquisition of land for managing wastes that are solid. The same can be done for agricultural waste and waste that is produced during mining. The administrator also has the authority to carry out projects of demonstration for testing and demonstrating techniques and methods that have been developed by people adopting this subsection of the Act.

Grants for Systems of Resource Recovery and Improvement of Facilities for Disposal of Wastes That Are Solid: This subsection of the act authorizes the administrator to make grants available to any state or municipality demonstrating systems for the recovery of resources and in the constructions of facilities that are new and improved for the purpose of disposing wastes. Any grants given based on this subsection will only be approved if it matches plans that meet specifications of subtitle D of the Act. There are also limitations to offering the grant and federal share shall not be more than 50%if a project serves an area. The administrator will also be responsible for developing regulations that will be used in the establishment of procedural guidelines for the award of grants. There are also additional limitations provided in this subsection such as the provision of grants for acquisition of land, provision of a satisfactory project plan to the administrator and other conditions or requirements the administrator asks of the pursuant. For a single state, not more than 15% of the funds allocated per fiscal year will be authorized for appropriation to carry out this section of the Act. Regulations on the manner in which this section will apply in grant application for a project, will be prescribed by the administrator and this will apply to one or more states.

In the year the act was passed the appropriation of funds to perform activities based on the act was not supposed to exceed 35 million dollars for every fiscal year following the enactment of the Act.

Solid Waste Cleanup on Federal Lands in Alaska: This shall be done with the presidents' directive as he or she sees appropriate to allow the executive department to conduct studies in collaboration with the of Alaska to determine the best suited procedures for the removal of solid waste that is existent in the federal lands of Alaska. After the study has been conducted, the president will then submit results of the study as desirable to the public works committee and that responsible for foreign commerce not later than one year after enactment of the solid waste utilization act of 1976. After a period of six months the president will be required to submit another report of the recommendations for administrative actions, procedures, and the legislation that is required in the implementation of procedures and implementing procedures from the study. The administrator of the EPA can also offer assistance that is either technical or financial for programs of research. This is done for the purpose of demonstrating the effectiveness in dealing water supply contamination that are used by the public (Sands, 2003). The programs for research authorized by the subsection of this act is to be designed the waste treatment management program of new castle county and this is done in collaboration with the EPA to develop control methods for contamination by leachate in landfills that have been abandoned and landfills that are present country wide. The research is meant to find solutions that are alternative to the actions for correction. The EPA administrator will also avail personnel include those from the laboratory responsible for research on solid or waste that is hazardous to provide assistance in conducting research. The administrator will also provide funds that will aid in protecting public water from further contamination while in the processes of conducting research. This is to cater for operation costs for the program of preventing dumping for the containment of leachate from landfills. In the event hazardous waste is released to the environment a cleanup process is undertaken and corrective measures are made. The procedures to follow have provisions in the amendment of 1980 and 1984 of the Act. The process of undertaking corrective action includes the evaluation of the and nature of the released waste. It also involves the development and implementation of measures to correct the problem in order to protect the public and the environment (Sands, 2003) (Figure 8.6).

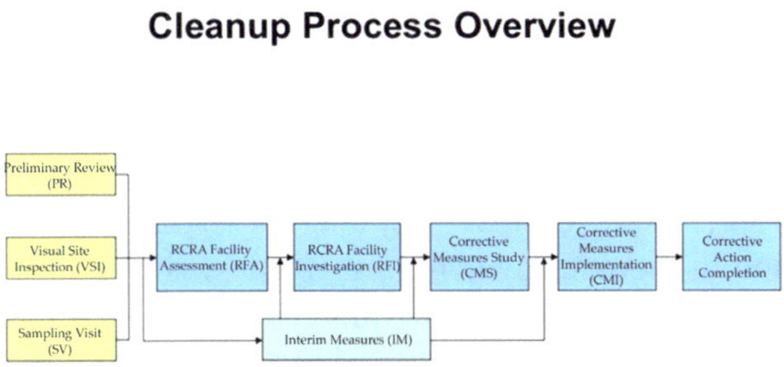

Figure 8.6: An overview of steps that can be followed in the clean-up process.

Source: https://dtsc-topock.com/resource-conservation-and-recovery-act.

8.1.2.9. Subtitle I – Underground Storage Tanks

The operation of underground storage tanks was added to the amendment of the resource and recovery act in the year 1984 in relation to waste that is either solid or hazardous. At this time 2.1 million tanks were subject to regulations by federal law and the EPA was front line in the closure of tanks that were found to be substandard. In 2009 600,000 underground storage tanks were approximately active at 223,000 sites and similarly subject to federal regulation.

Regulatory Requirements: It covers the storage of petroleum in tanks or substances that are considered and listed as hazardous and it also defines the nature and types of tanks that are permitted for use. This followed the establishment of a notification system for the purpose of tracking UST status, programs for regulation by the EPA but their administration has remained the responsibility of agencies in the United States. There are also regulatory standards for ground monitoring, double liners, the control of spill, the control of over fill and restrictions in relation to the disposal of untreatable waste that is hazardous (Ferrey, 2010).

The amendment of 1986 required those operating underground tanks to ensure corrective action when repair or removal is required or if it is deemed necessary to protect the environment or the health of humans. In the amendment provisions were made to create a trust fund that would cater for the cost of clean-up in the event there is leakage at sites where there are underground storage tanks in the case where those responsible for the leaks cannot be identified. This subtitle also recommend substitution with above ground storage tanks if possible.

8.1.2.10. Subtitle J – Medical Waste

This subtitle function to regulate the handling of medical waste in four states that is New York, Jersey, Connecticut, and Rhode Island but it expired in 1991. regulation of the management of medical waste is done by agencies in states related to health and the environment rather than the EPA. Other agencies include the center for disease prevention and control, occupational safety and administration and finally the food and drug administration (Ferrey, 2010).

8.1.3. Amendments and Legislation

Figure 8.7: Amendments and legislations of the resource recovery and conservation Act.

Source: https://19january2017snapshot.epa.gov/rcra/history-resource-conservation-and-recovery-act-rcra_.html.

8.1.3.1. Solid Waste Disposal Amendments of 1980

Several waste types have been exempted by Congress from being classified as wastes that are hazardous under the class C subtitle in the amendment of 1980 to the resource conservation and recovery Act. The amendment of 1980 on solid waste disposal classified some of the waste as special waste and they were not to be subjected to permitting requirements that were strict of subtitle C (Figure 8.7).

- Coal combustion residuals which are produced by electric power plants and other industries including fly ash, bottom ash; waste from slag and blue gas desulfurization wastes;
- Wastes from mines such as ore mines and mineral mines;
- Dust from cement kilns; and
- Fluids for drilling, water produced and waste that are produced from drilling oils or gases.

There are exemptions made and are referred to as either the Bevill or Bentsen exclusion which were previously temporary and awaiting further studies to be conducted to determine whether any of the wastes were considered as hazardous by the EPA.

8.1.3.2. Regulations Based on the Resource and Recovery Act

CCR regulations were established by the EPA in the year 2015 and this regulation was to aid in the restriction of the use of unlined ponds by coal powered plants. In 2018, the legislation was also modified by the Trump administration and this has been challenged and brought forward to the EPA for review by the courts of appeal for the United States for the District of Columbia Circuit. EPA afterwards published a proposed rule after a court decision was made in December 2019 and would be established in 2020 August which was the deadline given for establishments to stop the placement of ash to line fish ponds or impoundments. The proposal that was made also required more time to be given to establishments and this was up to the period of eight years to find other alternatives for managing wastes from ash before the closure of surface impoundments.

8.1.3.3. Superfund's

The superfund was created in the 1980s and is similarly referred as to the Comprehensive Environmental Response, Compensation, and Liability Act. It was meant to address to the problem of restitution of hazardous waste disposal sites that were abandoned, by the establishment of legal liability as well as a fund to facilitate the activities involved in cleanup process. The fund applies to sites that are contaminated while the RCRA major focus was on the control of particular waste generation and management and similarly to the CERCLA provides the resources needed for contamination site cleanup that had previously been done.

8.1.3.4. Hazardous and Solid Waste Amendments of 1984

The scope of the resource and recovery act of 1984, was broadened by the congress by enacting the amendments for hazardous and solid waste. Environmental laws on waste management were strengthened by the amendment which covered sections dealing with small quantity generators of wastes that are hazardous, incinerators, and the closure of landfills considered as substandard.

8.1.3.5. Land Disposal Program Flexibility Act of 1996

The land disposal program flexibility act of 1996 gave way for some procedure pliability for some waste to be disposed on land. For example, waste that is sent to industrial wastewater treatment, is not subject to restrictions of land disposal or if it is treated in a facility that has zero discharge.

8.1.4. Treatment, Storage and Disposal Facility Permits

Management of hazardous waste in facilities that is the waste treatment, storage, and disposal under subtitle C of the resource conservation and recovery act which requires a facility to have seek a permit before the commencement of its operation in some cases facilities continue to work under an interim status. Provisions related to acquiring an interim status is found under the 40 CFR part 265. For treatment, storage, and disposal facilities the requirements for permits appear in the 40 CFR parts 264 and 270 is for the management of waste that is considered to be hazardous in different units for example container storage areas, tanks, surface impoundments, waste piles, units for land treatment, landfills, incinerators, buildings for containment and drip pads. The unit-specific that gives operational and permitting requirements are further described in further detail in 40 CFR part 264, subparts J through D (Figure 8.8).

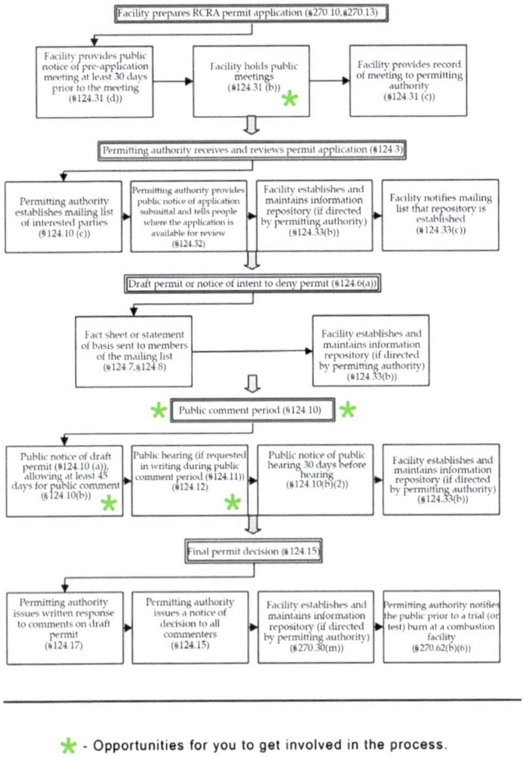

Figure 8.8: RCRA process of acquiring an operating permit.

Source: https://leanweb.org/tools-resources/louisiana-citizens-guide-environmental-engagement/chapter-2-public-participation-government/2-4-permit-process/.

8.2. EU WASTE FRAMEWORK DIRECTIVE

The Waste Framework Directive, also known as Directive 2008/98/EC on waste and the repeal of certain directives issued by the European Parliament as well as Council on November 19, 2008. The Directive repealed Rule 2006/12/EC (a codified version of the revised Directive 75/442/EEC) issued by the European Parliament and Council on April 5, 2006, and the Hazardous Waste Directive 91/689/ EEC, plus Waste "Oil Directive 75/439/ EEC." This provides a general structure for waste management needs and establishes a basic waste managing definition for the EU. Regulation (EC)

Number. 1013/2006, issued by the European Parliament and Council on June 14, 2006, on waste transportation.

The regulation specifies the conditions under which waste can be transported between nations. Decision 2000/532/EC presenting the waste list. The decision established a waste classification system, including the distinction between hazardous waste and non-hazardous waste. It's closely related to the catalog of main features that make waste hazardous, captured in Appendix III of the Waste Structure Directive. New legislative updates around the number of waste and hazardousness elements are also (applicable as of 1st June 2015).

8.3. BASEL CONVENTION

The Plenipotentiary Conference adopted the Basel Convention on Control of Transboundary Transportation of Hazardous Wastes as well as their Disposal in Basel, Switzerland, on March 22, 1989, in reaction to strong public protests after the discovery in Africa and other regions in the 1980s, with respect to deposits of toxic waste imported by developing countries from abroad. The awakening of environmental awareness and the corresponding constricting of environmental regulations for industrialized worlds during the 1970s and 1980s, leading the public to increasingly object to disposal of hazardous waste under the so-called NIMBY (not in the backyard) program and to boycott the public, disposal costs have further risen.

This, in turn, has led some individuals to seek affordable hazardous waste disposal options in Eastern Europe and developing worlds, while the knowledge of the environment in Eastern Europe and the developing world is far from sufficient, lacking regulations and enforcement mechanisms. Against this background, the Basel Convention was ratified during the late 1980s, while its adoption was focused on combating the so-called "toxic trade." The convention was enacted in 1992. The primary goal of the Basel Convention is protecting human health and environment from the harmful impacts of hazardous waste.

9

Climate Change

CONTENTS

9.1. INTRODUCTION

Climate change is an existential threat and the Kyoto Protocol and the Paris Agreement were developed to manage it. Case studies of national legislation are available, as well as the challenges in legislating climate change.

9.2. KYOTO PROTOCOL

The "Kyoto Protocol" signed in Kyoto, Japan in 1997 is the main mechanism for the international community to take steps towards reducing greenhouse gas emissions. "Kyoto Protocol" has been succeeded by the "Paris Agreement "which was reached in December 2015. "Kyoto Protocol" came into effect in February 2005, instructing industrialized countries that have passed the agreement to reduce their six major greenhouse gas (GHG), with the main emission contributors being carbon dioxide, nitrous oxide and methane. The "Kyoto Protocol" is an important first step. This is the most widely agreed environmental agreement ever. This agreement marks the international community's willingness to recognize and take action to combat climate change.

According to the protocol, every industrialized country has group of binding greenhouse gas discharge target to minimize emissions to below 1990 by 2012. Every country has different goals. Canada has set a 6% emission reduction target. Since the country has allowed emissions to surge since 1990, Canada must now cut back emissions to 30% below current levels. Even though the Manitoba government cannot become a signatory to Kyoto Protocol, on June 12th, 2008, the Manitoba Provincial Assembly ratification of Greenhouse Gas Emission Reduction Act and climate change, passed the Kyoto Protocol.

9.3. PARIS AGREEMENT

The Paris Agreement proposes a global structure to prevent dangerous climate change through limiting global warming levels to well below 2°C, and working hard to limit it to below 1.5°C. It also seeks to enhance the ability of countries to respond to the effects of climate change and further support their efforts. Paris Agreement is the first legally binding universal global climate change agreement adopted during the Paris Climate Conference (COP21) of December 2015. The EU along with its member states are roughly 190 parties signed to Paris Agreement. The European Union (EU) formally approved the agreement on October 5, 2016, thereby making it effective on November 4, 2016.

In order for the agreement to take effect, 55 countries, which account for at least 55% of universal emissions, must deposit approvals. The Paris Agreement is the passage between modern-day policies and climate neutrality before the turn of the century. Governments have reached a long-term objective of keeping the global average temperatures rise below 2°C above pre-industrial levels; aiming to limit the temperature rise to less than 1.5°C, as this will greatly reduce risks and effects of climate change; recognizing that global emissions need to peak immediately, recognizing that developing countries will take longer; and then reducing emissions quickly based on the best science available.

9.4. CASE STUDIES

9.4.1. New Locks in Albertkanaal in Flanders, Belgium

The Albert Canal in east Flanders is a significant economic channel, however, it has experienced a rare period of reduced water flow during which it is not suitable for commercial transportation. Due to climate change, the Meuse River Basin, including the canal, is expected to experience more regular and longer durations of reduced water flow. This may have an economic impact. The Albert Canal attaches the industrial area around Liege with the port of Antwerp. Boats can cross both sides of the canal: through the Scheldt River to the Netherlands, and through the Meuse river to France. For Belgium, this is an important economic channel, with a total annual flow of 40 million tons.

This also helps to reduce the amount of trucks moving in the highway by about 6,000 trucks per day, and reduce air pollution and traffic congestion. In some presently rare cases, the emission of the Meuse River is insufficient to feed all canals of Flanders and the Netherlands. During these times, the reduced water levels of the Albert Canal must be reduced, which means that draughts allow ships to pass through, making inland navigation an unappealing means of transportation.

9.4.2. Barcelona Trees Tempering the Mediterranean City Climate, Spain

The main climate change issues facing Barcelona include rising temperatures, reduced rainfall, and increased extreme events like droughts and heatwaves. High urban densities of Barcelona will enhance the heat island impact. Barcelona is committed to being a global model for sustainable cities that respond to urban development challenges tied to climate change. For a number of years, Barcelona has been dedicated to planting and management of trees. Trees can cool the urban climate in two different ways. The sunlight and transpiration reflected by leaves lower the temperature, while the shade lowers the surface temperatures and protects individuals from the hot sun, particularly during the hottest months.

In addition, trees can help reduce rainwater runoff, thereby preventing local flooding. In addition to climate-related benefits, urban trees can provide the following common benefits: removal of air pollutants, storage of carbon, reduction of noise pollution, regulation of humidity and balanced water circulation, the establishment of ecological links, offering habitat for city biodiversity and developing attractive city landscapes.

9.5. CHALLENGES IN LEGISLATING CLIMATE CHANGE

Climate change is among the greatest challenges of modern times. Climate change is presently happening: temperatures are increasing, droughts, and wildfires are starting to appear more regularly, rainfall patterns are changing, glaciers, and snow melting, and the global average sea levels rising. To combat climate change, it's important to reduce or prevent emissions related to human activities.

Fluorinated gas has a high global warming potential. Everyone knows the negative impacts of carbon dioxide (CO_2), however, few people recognize that the global warming prospects of fluorinated gases (F-gas) is about one-third more than CO_2. Supervision by the EU in conjunction with the European Economic Area can be helpful.

Discussions about rising global warming temperatures can be confusing because different reports might use different baselines to measure the magnitude of the increase. The two most popularly used benchmarks are (a) pre-industrial circumstances, the temperature of which is based on a 30-year average in 1850 and (b) an average level between 1980 and 1999. The latter is the context for most policy deliberations on aspects to establish limits for global warming.

The latter is the benchmark used in the IPCC report and is 0.5°C higher than the pre-industrial benchmark. A temperature 2°C higher than the pre-industrial temperature level is equivalent to about 1.5°C higher than the temperature from 1980 to 1999, and about 1.2°C higher than the average temperature in 2007 (smoothed to allow yearly changes), because about 0.25°C increased from 1990 (the midpoint of 1980 to 1999) to 2007.

10

Problems/Challenges in Implementation of Environmental Laws

CONTENTS

10.1. INTRODUCTION

A significant step that is taken in ensuring that the environment is protected is has been the making of laws. But if the laws have been made and are not enforced then there is no need for them being present. The enforcement of these laws is what drive that causes the measures to be taken in order for the environment to be protected from all the possible dangers that it may face. The implementation of the various environmental laws has faced a variety of challenges. The challenges may vary depending on the state. For instance, the challenges faced in developing countries may not be the same as the ones that are faced in developed countries. A variety of factors come into play. The main reason for the implementation of the environmental laws is to ensure that the environment is maintained and protected so that the generations to come are able to enjoy a clean and healthy environment.

The major challenge that may face the implementation of the environmental laws is the lack of understanding from the government officials and also the citizens of the state. This lack of understanding will cause a reluctance in the implementation of these laws as people do not know to the extent in which the laws may be useful in improving the environment. When we talk of environmental laws, there are a variety depending on the area of the environment. For instance, with regards to water there is the Clean Water Act (CWA). There are other environmental laws which include the Environmental Quality Act, Fisheries Act, Land Conservation Act, National Forestry Act, Plant Quarantine Act, Protection of Wildlife Act, Radioactive Substance Act, Sewage Services Act, and the Water Enactment Act among others.

All these laws when enforced, will play a significant role in improving and maintaining the environment. Most of these acts are under the EPA. This implies that the EPA is the body in charge of the implementation of these laws. If the environmental laws put in place have failed, then one should be inclined to look into the implementation of the laws in place. In some cases, the laws might have been implemented but in the wrong manner. Thus they are poorly implemented. This may be attributed to the fact that most of the legislators put more emphasis on making the environmental laws rather than enforcing them. The other issue may be the fact that most of these responsibilities are placed on government officials.

This means that even though the environment may be suffering, the suffering may be in silence as most of the responsibilities are placed on the government officials and they tend to brush of matters concerning the environment. In most cases, drastic measures are taken when there are disasters or when there are contaminations threatening the lives of individuals. In this article, we get to discuss the various problems at affect the implementation of these laws fully. For one to know the level in which the laws have not been implemented, assessments are conducted to evaluate. Some of the barriers faced in implementing the laws include the following:

10.2. LACK OF POLITICAL WILL AND STAKEHOLDER SUPPORT

Maintaining political and stakeholders' support in implement environmental flow provision. This can be stated as lack of political will and stakeholder support. The public alone cannot be involved in the implementation of these laws. Also changes in policy may not lead to the implementation of these laws. Both political and public support is required for the implementation of these laws. The major reason why there may be a reluctance of the political sector to implement these laws. This may be attributed to the lack of knowledge and understanding on the importance of these laws. With the lack of understanding of these laws, they may not be able to identify the significance of these laws. With the lack of understanding, they will not support the implementation of these laws.

The other reason why the political sector may not support the implementation of these laws is due to lack of proper strategies that are present to facilitate the implementation of these laws. Support from the political side is very important. This is because it allows for the setting of strategic direction, securing planning resources and championing environmental requirement which will result to the implementation of these laws. It is therefore important for the legislator to ensure that they properly educate the political leaders on the environmental laws.

The other challenge that may lead to the political sector not implementing the laws is lack of coordination between the various sectors in the environment. For instance, when handing the environment, sectors such as earth, water, and oceans work together to ensure that the environment is well protected. In the case of clean water, one has to ensure that the land is clean to ensure that the land is clean so that the water is clean. This means that the land, water, and agricultural sectors of the government need to work together so that the environmental laws become a success.

The challenges presented by the political areas are monumental. For instance, when it comes to managing the environmental flows, it may mean that either water should be removed from initial uses or constraints be placed on the future water uses. This may affect the agricultural, real estate, hydropower or the industrial sectors may be affected by the changes wield potential political power and it may lead to the resistance in reforming water management. The major needs for social and economic requirements for water resources may lead to the political bias against the implementation of these laws.

For some of the political areas, they are hesitant in implementing these laws as they lack the necessary resources that may be useful in the implementation of the laws. For some of these laws, they require programs to be put in place so that the various environmental acts become effective. Some of the states may lack the needed funds to support these programs being initiated. The other reason of the failure of the implementation of the environmental laws is lack of support from the stakeholders. Stakeholders play a major role in the implementation of these laws in terms of funds.

Most of the environmental laws are implemented by the EPA. It is the body in place in most of the countries that were present in the world summits. The EPA has been known to implement most of the environmental laws in place. For the EPA, they are unable to implement the environmental laws if they lack support from the related stakeholders. The stakeholders help the EPA through the provision of funds which is utilized in the enforcing of these laws. The stakeholders play a crucial in ensuring that the boards run and can be in place for many years. The government may also influence the decision of the stakeholders. If the government gives support to the EPA then it will encourage the stakeholders to form alliances with the required programs so that they can ensure that they are implemented successfully.

The implementation of ground actions by the government will ensure that the environmental laws are implemented fully. They are more like test runs. If they are successful, then it will imply that the environmental laws will be effectively implemented and used. Most of the laws are undergoing reformation. For instance, in South Africa there are efforts being made to reform their Water Act as per the European Union's Water Framework Directive. These efforts have demonstrated that is one thing to pass the ambitious laws and policies and a different thing to implement the ground actions ensuring that they are effective. In the implementation stage the policy reforms may be faced with challenges which include politically changing realities. In some instances, there may be resistance that may be from the affected areas more so in politically unpopular decisions which includes increased water bills.

Therefore, the agreements made concerning water management may be face some challenge in the case where signatory states do not have common goals. This is as per the case study that was conducted in lower Mekong River basin.

10.3. INSUFFICIENT RESOURCES AND CAPACITY

Resources are among the major factors which affect the implementation of the environmental laws. Resources are a variety of things. The major resource may be funds which is crucial in ensuring that most of the projects in place are achieved. The implementation of these laws cannot be achieved if there are no strong institutions with sufficient resources as well as the capacity to conduct them. The case studies that have been conducted have shown that the limited resources have causes some constraint in achieving an implementation of these laws.

In the case of resources, gathering resources has been a challenge. For instance, in the case of handling hazardous wastes the major challenge has been getting the funds to facilitate the cleanup projects. Initially, the individuals who were responsible for the wastes were in charge of paying up of for the cleanup process. Though this was advantageous to some extent, the problem was that later on there were many sites that were discovered limiting the available funds. The Superfund therefore had to look for other ways in which they can pay for the cleanup. The other challenge that they encountered was that in some states, the government cut off some of the funds going to the Superfund causing a lot of projects to be stalled.

The case study above just goes to show the importance of foods in implementation of environmental laws. In the case of the Superfund the environmental law that was to be implemented is the cleaning up of identified waste sites and is under the CERCLA Act. The resources are mostly funds. In some other cases the resources may be facilities among other things. In some other case studies, it was discovered that in light of doing operational rules that affect the environment, even a single dam or river need relevant technical as well as institutional capacity. Implying that if the environmental laws are to be implemented in a larger scale, there will be great need for resources.

If the challenges were to be compounded, then it would involve the provision of adequate finances will be required to facilitate the building and maintenance of effective environmental laws management programs. It is also important as it is utilized in the environmental laws assessment to incentives for the flow re-allocation to the long term monitoring as well as adaptive management. This substantiates the importance of adequate resources during the implementation of environmental laws. For the case of a comprehensive framework in the implementing of environmental laws need relevant laws, policies, procedures, regulations, and institutions to be in place so that they can allow for the implementation of laws more so for the ones dealing with environmental management.

In the context above the implementation of the environmental laws effectively requires to conditions. They include effective water management and allocation policy and institutional framework as well as the recognition of the environmental flow requirements within the framework. They are both required to ensure that they work. A single condition on it's on cannot work on its own. It needs the rest to ensure that the laws are being implemented. Take for instance in the following require main mechanism and capacities that are related to these conditions.

For instance, the policy and legislation have both there is the water management framework and the environmental law framework. The water management framework involves a clear legal basis is in place for issuing and regulation water use, rights, and licenses. There are also national laws and policies which provide a framework for national and basin water planning. The environmental framework involves environmental flows being legally recognized as an important requirement. There is also a legal status or mandate which allows setting and the regulation of target water flows. There is also the assessment modeling which involves technical capacity is in existence used in building plan and is funded by a long-term hydrological data acquisition management. The environmental framework involves a methodology used in determining environment flow needs are established.

For the successful implementation of the constituent elements of the framework may present a relevant challenge. The environmental law policies can be enacted by either comprehensive or incremental reform. The comprehensive reforms lead to a change in the water resource management framework this gives off an opportunity to strengthen the water management institution and also incorporate environmental laws requirement within them. (Bodansky, 2010)

In the implementation of the environmental laws needs the building of frameworks. They require significant managerial, financial, and technical capacities. They are however they are not limited to the ability to assess resource availability and environmental conditions, they also engage and facilitate stakeholder, undertake environmental flows assessment and they also enforce water allocation and reservoir release schedules among other things. The implementation of the environmental laws involves implementation of environmental management which may require a lot of money. There is need of the resources to be provided to ensure that the laws are implemented.

The financial resources are required to ensure that environmental assessment, building of data bases and the capacities required to build them. A good case of how the implementation of environmental laws failed was in the Alaska environmental law protection Programme and it is attributed to the lack of adequate funding. There should also be a capacity to allow the implementation of the laws. Availability of a capacity will allow the implementation of several laws.

10.4. INSTITUTIONAL BARRIERS AND CONFLICTS OF INTEREST

The environmental laws are a relater interdisciplinary and inter-sectoral. There has been an identification that the understanding, incorporation as well as the improving legal and institutional frameworks as a major challenge in the implementation of the environmental laws solutions. The agencies planning and managing hydropower, agriculture, land use, urban, and industrial development as well as the natural resources involves the use of legal authorities. The lack of understanding present concerning the interdependence between the various downstream water needs and aquifers which clearly demonstrate the institutional barriers present in government agencies managing them.

The lack of understanding causes conflict among the legislators which may lead to the environmental laws not being implemented. The lack of understanding in countries whose political as well as the academic institutions do not cooperate may lead to a setback in the developing scientific as well as decision support tools required in the setting of these laws. The areas of conflicts may also arise in the parties involved in the implementing agencies. This present a significant challenge in the implementation of the laws. There is need for institutions to allow the implementation of these laws. For instance, in the managing of hazardous waste sites, there is need for institutional facilities that will be needed in the managing of wastes. The challenge has been that there is an insufficient number of facilities that will be need cleaning these areas. This is one way in which institutional barriers present itself.

The institutional barrier become poignant when interstate agreements are vehicles for environmental laws recognition. There was an interstate agreement was signed and this was because it was driven by the lack of overarching institution having a legal mandate to facilitate the management of resources. The lack of political authority may contribute to the lack of implementation of the environmental laws.

10.5. LACK OF PROPER STRATEGY

The strategy is important in ensuring that the laws are implemented. Strategies involve an outline of steps that will be taken to ensure that the laws are implemented. There are universally present laws that are used in implementing these laws. However, these laws may not apply to all the states. For the different states, there is a unique strategy that will effectively work for it. In the case of poor implementation of these laws or lack of implementation of these laws, it majorly revolves around the strategy that has been employed. If the strategy that has been employed is not clear, then the laws may not be successfully implemented.

Also, if the employed strategies do not meet the task at hand, then it is most likely laws will not be implemented successfully. A poor strategy will also contribute to political sectors being reluctant in implementing some of these laws. Usually, before the laws are implemented, first the strategy that will be used in implementing this law are presented before the political sector and if they approve the given strategy then the law is implemented. It is therefore the work of the legislators to ensure that the strategy to be employed is appropriate. The strategies may involve the systems that should be involved when these laws are implemented.

10.6. CONSUMER AND INDUSTRIAL PRESSURES

Another reason why the environmental laws have not been implemented is the demand from the population. For instance, there are laws that have been set concerning forests which restrict the cutting of trees but most of these laws have not been implemented because of the demand from the population. This has seen most of the trees to be cut leading to the large of forests that have been totally destroyed though there are laws that have been set to protect these forests. In some cases, the customer demand may affect the implementation of these environmental laws. Some of the government sectors work hand in hand with some of the industrial sectors. To some extent, they may allow the needs of the population overcome the need for them making appropriate decision of implementing these laws. This causes a major setback in the implementation of the environmental laws.

There is need for the legislators to make the government officials understanding the implications of allowing the environmental needs overcoming the need to enforce laws that will protect the environment.

To some extent, corruption in the organizations involved in the implementation of the environmental laws. Some case studies have come to show that some of the environmental officials have stalled the projects that have been up. This to some extent has affected the implementation of laws as the working of these projects are ways in which the laws are being implemented.

Bibliography

1. Atapattu, S., (2007). Emerging Principles of International Environmental Law (pp. 1–77.). Netherlands: BRILL.
2. Banks, J., (2018). Importance Of Environmental Laws. Retrieved from statonlaw.net: https://www.statonlaw.net/importance-of-environmental-laws/ (accessed on 11 July 2020).
3. Bearden, D. M., (2012). Comprehensive environmental response, of superfund cleanup authorities and compensation, and liability act: A summary related provisions of the Act. CRS Report for Congress, 1–40.
4. Beder, S., (2013). Environmental Principles and Policies: An Interdisciplinary Introduction (pp. 1–69). London: Routledge.
5. Bodansky, D., (2010). The Art and Craft of International Environmental Law. London: Harvard University Press.
6. Bos, R., (2016). Manual on the Human Rights to Safe Drinking Water and Sanitation for Practitioners (pp. 3–65). London: IWA Publishing.
7. Bosselmann, K., (2016). The Principle of Sustainability: Transforming Law and Governance (pp. 1–54). London: Taylor & Francis.
8. Carter, T. W., (2006). Safe Drinking Water Act and Its Interpretation (pp. 23–125.). New York: Nova Publishers.
9. Chay K., & Greenstone, M., (2003). Air Quality, Infant Mortality, and the Clean Air Act of 1970. Massachusett, https://www.nber.org/papers/w10053.pdf: NATIONAL BUREAU OF ECONOMIC RESEARCH (accessed on 11 July 2020).
10. Dave, F., & Allison, F., (2009). Overview of the Resource Conservation and Recovery Act ("RCRA") and Comprehensive Environmental

Response, Compensation, and Liability Act ("CERCLA"). Pennsylvania: Environmental Law Institute.

11. Doub, J. P., (2016). The Endangered Species Act: History, Implementation, Successes, and Controversies (pp. 1–73). Florida: CRC Press.
12. Ferrey, S., (2010). Environmental Law: Examples and Explanations (p. 671). Pennsylvania: Aspen Publishers Online.
13. Fisher, E. C., (2017). Environmental Law: A Very Short Introduction (pp. 1–81). London: Oxford University Press.
14. Foreman, P., (2002). Endangered Species: Issues and Analyses (pp. 1–55). New York: Nova Publishers.
15. Glavan, M., (2018). Water Challenges of an Urbanizing World (pp. 13–87). Berlin: BoD-Books on Demand.
16. Houck, O. A., (2002). The Clean Water Act TMDL Program: Law, Policy, and Implementation (pp. 11–99). New York: Environmental Law Institute.
17. Jonathan, M. D., & Joseph, M. N., (2011). An Interactive History of the Clean Air Act: Scientific and Policy Perspectives (pp. 7–141). New York: Elsevier.
18. Kate, C. S., & Richard, K. L., (2020). Clean Air Act: A Summary of the Act and Its Major Requirements. New York: Congressional Research Service.
19. Koivurova, T., (2013). Introduction to International Environmental Law (pp. 1–53). London: Routledge.
20. Lazarus, R. J., (2008). The Making of Environmental Law (pp. 1–43). Chicago: University of Chicago Press.
21. Leonard, B., (2011). Identifying Waters Protected by the Clean Water Act (CWA): Draft Guidance (pp. 1–34). New York: DIANE Publishing.
22. Louka, E., (2006). International Environmental Law: Fairness, Effectiveness, and World Order. New York: Cambridge University Press.
23. Macrory, R., (2008). Regulation, Enforcement and Governance of Environmental Law (pp. 11–35). Cameron: Cameron May.
24. McAllister, L., (2008). Making Law Matter: Environmental Protection and Legal Institutions in Brazil (pp. 1–57). California: Stanford University Press.

25. Mittal, A. K., (2010). Clean Water Act: Long-standing Issues Impact EPA¿s and States¿ Efforts; Testimony Before the Committee on Transportation and Infrastructure, U. S. House of Rep (p. 18). New York: DIANE Publishing.
26. Nazzaro, R., (2010). Endangered Species Act: Many Recommendations Have Been Implemented, But Some Issues Remain Unresolved (p. 215). New York: DIANE Publishing.
27. Pontius, F., (2003). Drinking Water Regulation and Health (pp. 19–45). Florida: John Wiley & Sons.
28. Republic of South Africa, (2020). Constitution of the Republic of South Africa Act No. 108 of 1996.Available at: https://www.gov.za/sites/default/files/images/a108-96.pdf
29. Rostron, J., (2013). Environmental Law for The Built Environment (p. 424). London: Routledge.
30. Sadeleer, N. D., (2002). Environmental Principles: From Political Slogans to Legal Rules (pp. 1–21). London: Oxford University Press.
31. Sancin, V., (2012). International Environmental Law: Contemporary Concerns and Challenges: Papers Presented at the First Contemporary Challenges of International Environmental Law Conference (pp. 9–85). Ljubljana, New Delhi: Masa Kovic Dine.
32. Sands, P., (2003). Principles of International Environmental Law (pp. 3–72). London: Cambridge University Press.
33. Shogren, J. F., (2013). Private Property and the Endangered Species Act: Saving Habitats, Protecting Homes (pp. 8–93). Texas: University of Texas Press.
34. Stephenson, J. B., (2005). Clean Water Act: Improved Resource Planning Would Help EPA Better Respond to Changing Needs and Fiscal Constraints (pp. 3–44). New York: DIANE Publishing.
35. Tiemann, M., (2011). Safe Drinking Water Act (SDWA): A Summary of the Act and Its Major Requirements (pp. 1–14). California: DIANE Publishing.
36. Trimble, D. C., (2011). Safe Drinking Water Act (SDWA): EPA Should Improve Implementation of Requirements on Whether to Regulate Additional Contaminants (pp. 92–139). California: DIANE Publishing.
37. The United States Environmental Protection Agency, (2007). The Plain English Guide to the Clean Air Act. North Carolina: Office of Air Quality Planning and Standards Research Triangle Park.

38. The United States Environmental Protection Agency, (2014). Resource Conservation and Recovery Act (p. 242). Pennsylvania, The U.S. Environmental Protection Agency Office of Resource Conservation and Recovery Program Management, Communications, and Analysis Office.
39. Sullins, T. A., (2001). ESA, Endangered Species Act (pp. 1–84). New York: American Bar Association.
40. The UN Sub-Commission on Human Rights, (2003). E/CN.4/Sub.2/2003/12/Rev.2. Norms on the Responsibilities of Transnational Corporations and Other Business Enterprises with regard to Human Rights. UN Economic and Social Council. Geneva, Switzerland.
41. Welker-Hood, K., Barbara, G., & John, S., (2011). The Clean Air Act. A Proven Tool for Healthy Air, 40.

INDEX

R

S

T

U

V

W